The Laws of Thermodynamics: A Very Short Introduction

VERY SHORT INTRODUCTIONS are for anyone wanting a stimulating and accessible way into a new subject. They are written by experts, and have been translated into more than 45 different languages.

The series began in 1995, and now covers a wide variety of topics in every discipline. The VSI library now contains over 500 volumes—a Very Short Introduction to everything from Psychology and Philosophy of Science to American History and Relativity—and continues to grow in every subject area.

Titles in the series include the following:

Peter Atkins

THE LAWS OF
THERMODYNAMICS

A Very Short Introduction

OXFORD
UNIVERSITY PRESS

OXFORD
UNIVERSITY PRESS

Great Clarendon Street, Oxford OX2 6DP

Oxford University Press is a department of the University of Oxford.
It furthers the University's objective of excellence in research, scholarship,
and education by publishing worldwide in

Oxford New York

Auckland Cape Town Dar es Salaam Hong Kong Karachi
Kuala Lumpur Madrid Melbourne Mexico City Nairobi
New Delhi Shanghai Taipei Toronto

With offices in

Argentina Austria Brazil Chile Czech Republic France Greece
Guatemala Hungary Italy Japan Poland Portugal Singapore
South Korea Switzerland Thailand Turkey Ukraine Vietnam

Oxford is a registered trade mark of Oxford University Press
in the UK and in certain other countries

Published in the United States
by Oxford University Press Inc., New York

© Peter Atkins 2010

First published in hardback as
Four Laws that Drive the Universe 2007
First published as a Very Short Introduction 2010

British Library Cataloguing in Publication Data
Data available

Library of Congress Cataloging in Publication Data
Data available

Typeset by SPI Publisher Services, Pondicherry, India
Printed and bound by
CPI Group (UK) Ltd, Croydon, CRO 4YY

ISBN 978–0–19–957219–9

Contents

Preface

Among the hundreds of laws that describe the universe, there lurks a mighty handful. These are the laws of thermodynamics, which summarize the properties of energy and its transformation from one form to another. I hesitated to include the word 'thermodynamics' in the original title of this little introduction to this boundlessly important and fascinating aspect of nature, hoping that you would read at least this far, for the word does not suggest a light read. And, indeed, I cannot pretend that it will be a light read. When in due course, however, you emerge from the other end of this slim volume, with your brain more sinewy and exercised, you will have a profound understanding of the role of energy in the world. In short, you will know what drives the universe.

Do not think that thermodynamics is only about steam engines: it is about almost everything. The concepts did indeed emerge during the nineteenth century when steam was the hot topic of the day, but as the laws of thermodynamics became formulated and their ramifications explored it became clear that the subject could touch an enormously wide range of phenomena, from the efficiency of heat engines, heat pumps, and refrigerators, taking in chemistry on the way, and reaching as far as the processes of life. We shall travel across that range in the pages that follow.

The mighty handful consists of four laws, with the numbering starting inconveniently at zero and ending at three. The first two laws (the 'zeroth' and the 'first') introduce two familiar but nevertheless enigmatic properties, the temperature and the energy. The third of the four (the 'second law') introduces what many take to be an even more elusive property, the entropy, but which I hope to show is easier to comprehend than the seemingly more familiar properties of temperature and energy. The second law is one of the all-time great laws of science, for it illuminates why anything—anything from the cooling of hot matter to the formulation of a thought—happens at all. The fourth of the laws (the 'third law') has a more technical role, but rounds out the structure of the subject and both enables and foils its applications. Although the third law establishes a barrier that prevents us from reaching the absolute zero of temperature, of becoming absolutely cold, we shall see that there is a bizarre and attainable mirror world that lies below zero.

Thermodynamics grew from observations on bulk matter—as bulky as steam engines, in some cases—and became established before many scientists were confident that atoms were more than mere accounting devices. The subject is immeasurably enriched, however, if the observation-based formulation of thermodynamics is interpreted in terms of atoms and molecules. In this account we consider first the observational aspects of each law, then dive below the surface of bulk matter and discover the illumination that comes from the interpretation of the laws in terms of concepts that inhabit the underworld of atoms.

In conclusion, and before you roll up the sleeves of your mind and get on with the business of understanding the workings of the universe, I must thank Sir John Rowlinson for commenting in detail on two drafts of the manuscript: his scholarly advice was enormously helpful. If errors remain, they will no doubt be traced to where I disagreed with him.

List of illustrations

Chapter 1
The zeroth law

The concept of temperature

The zeroth law is an afterthought. Although it had long been known that such a law was essential to the logical structure of thermodynamics, it was not dignified with a name and number until early in the twentieth century. By then, the first and second laws had become so firmly established that there was no hope of going back and renumbering them. As will become apparent, each law provides an experimental foundation for the introduction of a thermodynamic property. The zeroth law establishes the meaning of what is perhaps the most familiar but is in fact the most enigmatic of these properties: temperature.

Thermodynamics, like much of the rest of science, takes terms with an everyday meaning and sharpens them—some would say, hijacks them—so that they take on an exact and unambiguous meaning. We shall see that happening throughout this introduction to thermodynamics. It starts as soon as we enter its doors. The part of the universe that is at the centre of attention in thermodynamics is called the *system*. A system may be a block of iron, a beaker of water, an engine, a human body. It may even be a circumscribed part of each of those entities. The rest of the universe is called the *surroundings*. The surroundings are where we stand to make observations on the system and infer its properties. Quite often, the actual surroundings consist of a water bath maintained at constant temperature, but that is a more controllable approximation to the true surroundings, the rest of the world. The system and its surroundings jointly make up the *universe*. Whereas for us the universe is everything, for a less profligate thermodynamicist it might consist of a beaker of water (the system) immersed in a water bath (the surroundings).

A system is defined by its boundary. If matter can be added to or removed from the system, then it is said to be *open*. A bucket, or more refinedly an open flask, is an example, because we can just shovel in material. A system with a boundary that is impervious to matter is called *closed*. A sealed bottle is a closed system. A system with a boundary that is impervious to everything in the sense that the system remains unchanged regardless of anything that happens in the surroundings is called *isolated*. A stoppered vacuum flask of hot coffee is a good approximation to an isolated system.

The properties of a system depend on the prevailing conditions. For instance, the pressure of a gas depends on the volume it occupies, and we can observe the effect of changing that volume if the system has flexible walls. 'Flexible walls' is best thought of as meaning that the boundary of the system is rigid everywhere except for a patch—a piston—that can move in and out. Think of a bicycle pump with your finger sealing the orifice.

Properties are divided into two classes. An *extensive property* depends on the quantity of matter in the system—its extent. The mass of a system is an extensive property; so is its volume. Thus, 2 kg of iron occupies twice the volume of 1 kg of iron. An *intensive property* is independent of the amount of matter present. The temperature (whatever that is) and the density are examples. The temperature of water drawn from a thoroughly stirred hot tank is the same regardless of the size of the sample. The density of iron is 8.9 g cm^{-3} regardless of whether we have a 1 kg block or a 2 kg block. We shall meet many examples of both kinds of property as we unfold thermodynamics and it is helpful to keep the distinction in mind.

Introducing equilibrium

So much for these slightly dusty definitions. Now we shall use a piston—a movable patch in the boundary of a system—to introduce one important concept that will then be the basis for introducing the enigma of temperature and the zeroth law itself.

Suppose we have two closed systems, each with a piston on one side and pinned into place to make a rigid container (Figure 1). The two pistons are connected with a rigid rod so that as one moves out the other moves in. We release the pins on the piston. If the piston on the left drives the piston on the right into that system, we can infer that the pressure on the left was higher than that on the right, even though we have not made a direct measure of the two pressures. If the piston on the right won the battle, then we would infer that the pressure on the right was higher than that on the left. If nothing had happened when we released the pins, we would infer that the pressures of the two systems were the

1. If the gases in these two containers are at different pressures, when the pins holding the pistons are released, the pistons move one way or the other until the two pressures are the same. The two systems are then in mechanical equilibrium. If the pressures are the same to begin with, there is no movement of the pistons when the pins are withdrawn, for the two systems are already in mechanical equilibrium

same, whatever they might be. The technical expression for the condition arising from the equality of pressures is *mechanical equilibrium*. Thermodynamicists get very excited, or at least get very interested, when nothing happens, and this condition of equilibrium will grow in importance as we go through the laws.

We need one more aspect of mechanical equilibrium: it will seem trivial at this point, but establishes the analogy that will enable us to introduce the concept of temperature. Suppose the two systems, which we shall call A and B, are in mechanical equilibrium when they are brought together and the pins are released. That is, they have the same pressure. Now suppose we break the link between them and establish a link between system A and a third system, C, equipped with a piston. Suppose we observe no change: we infer that the systems A and C are in mechanical equilibrium and we can go on to say that they have the same pressure. Now suppose we break that link and put system C in mechanical contact with system B. Even without doing the experiment, we know what will happen: nothing. Because systems A and B have the same pressure, and A and C have the same pressure, we can be confident that systems C and B have the same pressure, and that pressure is a universal indicator of mechanical equilibrium.

Now we move from mechanics to thermodynamics and the world of the zeroth law. Suppose that system A has rigid walls made of metal and system B likewise. When we put the two systems in contact, they might undergo some kind of physical change. For instance, their pressures might change or we could see a change in colour through a peephole. In everyday language we would say that 'heat has flowed from one system to the other' and their properties have changed accordingly. Don't imagine, though, that we know what heat is yet: that mystery is an aspect of the first law, and we aren't even at the zeroth law yet.

It may be the case that no change occurs when the two systems are in contact even though they are made of metal. In that case we say

4

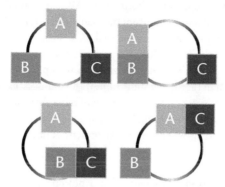

2. A representation of the zeroth law involving (top left) three systems that can be brought into thermal contact. If A is found to be in thermal equilibrium with B (top right), and B is in thermal equilibrium with C (bottom left), then we can be confident that C will be in thermal equilibrium with A if they are brought into contact (bottom right)

that the two systems are in *thermal equilibrium*. Now consider three systems (Figure 2), just as we did when talking about mechanical equilibrium. It is found that if A is put in contact with B and found to be in thermal equilibrium, and B is put in contact with C and found to be in thermal equilibrium, then when C is put in contact with A, it is always found that the two are in thermal equilibrium. This rather trite observation is the essential content of the *zeroth law of thermodynamics*:

if A is in thermal equilibrium with B, and B is in thermal equilibrium with C, then C will be in thermal equilibrium with A.

The zeroth law implies that just as the pressure is a physical property that enables us to anticipate when systems will be in mechanical equilibrium when brought together regardless of their composition and size, then there exists a property that enables us to anticipate when two systems will be in thermal equilibrium regardless of their composition and size: we call

5

this universal property the *temperature*. We can now summarize
the statement about the mutual thermal equilibrium of the three
systems simply by saying that they all have the same temperature.
We are not yet claiming that we know what temperature is, all we
are doing is recognizing that the zeroth law implies the existence
of a criterion of thermal equilibrium: if the temperatures of
two systems are the same, then they will be in thermal equilibrium
when put in contact through conducting walls and an observer
of the two systems will have the excitement of noting that nothing
changes.

We can now introduce two more contributions to the vocabulary
of thermodynamics. Rigid walls that permit changes of state when
closed systems are brought into contact—that is, in the language
of Chapter 2, permit the conduction of heat—are called
diathermic (from the Greek words for 'through' and 'warm').
Typically, diathermic walls are made of metal, but any conducting
material would do. Saucepans are diathermic vessels. If no change
occurs, then either the temperatures are the same or—if we know
that they are different—then the walls are classified as *adiabatic*
('impassable'). We can anticipate that walls are adiabatic if they
are thermally insulated, such as in a vacuum flask or if the system
is embedded in foamed polystyrene.

The zeroth law is the basis of the existence of a *thermometer,* a
device for measuring temperature. A thermometer is just a
special case of the system B that we talked about earlier. It is a
system with a property that might change when put in contact
with a system with diathermic walls. A typical thermometer
makes use of the thermal expansion of mercury or the change in
the electrical properties of material. Thus, if we have a system B
('the thermometer') and put it in thermal contact with A, and
find that the thermometer does not change, and then we put the
thermometer in contact with C and find that it still doesn't
change, then we can report that A and C are at the same
temperature.

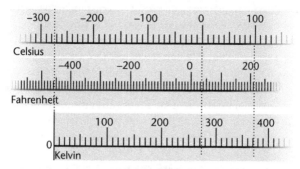

3. Three common temperature scales showing the relations between them. The vertical dotted line on the left shows the lowest achievable temperature; the two dotted lines on the right show the normal freezing and boiling points of water

There are several scales of temperature, and how they are established is fundamentally the domain of the second law (see Chapter 3). However, it would be too cumbersome to avoid referring to these scales until then, though formally that could be done, and everyone is aware of the Celsius (centigrade) and Fahrenheit scales. The Swedish astronomer Anders Celsius (1701–1744) after whom the former is named devised a scale on which water froze at 100° and boiled at 0°, the opposite of the current version of his scale (0°C and 100°C, respectively). The German instrument maker Daniel Fahrenheit (1686–1736) was the first to use mercury in a thermometer: he set 0° at the lowest temperature he could reach with a mixture of salt, ice, and water, and for 100° he chose his body temperature, a readily transportable but unreliable standard. On this scale water freezes at 32°F and boils at 212°F (Figure 3).

The temporary advantage of Fahrenheit's scale was that with the primitive technology of the time, negative values were rarely needed. As we shall see, however, there is an absolute zero of temperature, a zero that cannot be passed and where negative temperatures have no meaning except in a certain formal sense, not one that depends on the technology of the time (see Chapter 5). It is therefore natural to measure temperatures by setting 0 at this lowest attainable zero and to refer to such absolute temperatures as the *thermodynamic temperature*. Thermodynamic temperatures are denoted T, and whenever that symbol is used in this book, it means the absolute temperature with $T = 0$ corresponding to the lowest possible temperature. The most common scale of thermodynamic temperatures is the *Kelvin scale*, which uses degrees ('kelvins', K) of the same size as the Celsius scale. On this scale, water freezes at 273 K (that is, at 273 Celsius-sized degrees above absolute zero; the degree sign is not used on the Kelvin scale) and boils at 373 K. Put another way, the absolute zero of temperature lies at $-273°$C. Very occasionally you will come across the *Rankine scale*, in which absolute temperatures are expressed using degrees of the same size as Fahrenheit's.

The molecular world

In each of the first three chapters I shall introduce a property from the point of view of an external observer. Then I shall enrich our understanding by showing how that property is illuminated by thinking about what is going on *inside* the system. Speaking about the 'inside' of a system, its structure in terms of atoms and molecules, is alien to classical thermodynamics, but it adds deep insight, and science is all about insight.

Classical thermodynamics is the part of thermodynamics that emerged during the nineteenth century before everyone was fully convinced about the reality of atoms, and concerns relationships between bulk properties. You can do classical thermodynamics even if you don't believe in atoms. Towards the end of the nineteenth century, when most scientists accepted that atoms were real and not just an accounting device, there emerged the version of thermodynamics called *statistical thermodynamics*, which sought to account for the bulk properties of matter in terms of its constituent atoms. The 'statistical' part of the name comes from the fact that in the discussion of bulk properties we don't need to think about the behaviour of individual atoms but we do need to think about the average behaviour of myriad atoms. For instance, the pressure exerted by a gas arises from the impact of its molecules on the walls of the container; but to understand and calculate that pressure, we don't need to calculate the contribution of every single molecule: we can just look at the average of the storm of molecules on the walls. In short, whereas dynamics deals with the behaviour of individual bodies, *thermo*dynamics deals with the average behaviour of vast numbers of them.

The central concept of statistical thermodynamics as far as we are concerned in this chapter is an expression derived by Ludwig Boltzmann (1844–1906) towards the end of the nineteenth century. That was not long before he committed suicide, partly because he found intolerable the opposition to his ideas from colleagues who were not convinced about the reality of atoms. Just as the zeroth law introduces the concept of temperature from the viewpoint of bulk properties, so the expression that Boltzmann derived introduces it from the viewpoint of atoms, and illuminates its meaning.

To understand the nature of Boltzmann's expression, we need to know that an atom can exist with only certain energies. This is the domain of quantum mechanics, but we do not need any of that subject's details, only that single conclusion. At a given temperature—in the bulk sense—a collection of atoms consists of some in their lowest energy state (their 'ground state'), some in the next higher energy state, and so on, with populations that diminish in progressively higher energy states. When the populations of the states have settled down into their 'equilibrium' populations, and although atoms continue to

9

jump between energy levels there is no net change in the populations, it turns out that these populations can be calculated from a knowledge of the energies of the states and a single parameter, β (beta).

Another way of thinking about the problem is to think of a series of shelves fixed at different heights on a wall, the shelves representing the allowed energy states and their heights the allowed energies. The nature of these energies is immaterial: they may correspond, for instance, to the translational, rotational, or vibrational motion of molecules. Then we think of tossing balls (representing the molecules) at the shelves and noting where they land. It turns out that the most probable distribution of populations (the numbers of balls that land on each shelf) for a large number of throws, subject to the requirement that the total energy has a particular value, can be expressed in terms of that single parameter β.

The precise form of the distribution of the molecules over their allowed states, or the balls over the shelves, is called the *Boltzmann distribution*. This distribution is so important that it is important to see its form. To simplify matters, we shall express it in terms of the ratio of the population of a state of energy E to the population of the lowest state, of energy 0:

$$\frac{\text{Population of state of energy } E}{\text{Population of state of energy } 0} = e^{-\beta E}$$

We see that for states of progressively higher energy, the populations decrease exponentially: there are fewer balls on the high shelves than on the lower shelves. We also see that as the parameter β increases, then the relative population of a state of given energy decreases and the balls sink down on to the lower shelves. They retain their exponential distribution, with progressively fewer balls in the upper levels, but the populations die away more quickly with increasing energy.

When the Boltzmann distribution is used to calculate the properties of a collection of molecules, such as the pressure of a gaseous sample, it turns out that it can be identified with the reciprocal of the (absolute) temperature. Specifically, $\beta = 1/kT$, where k is a fundamental constant called *Boltzmann's constant*. To bring β into line with the Kelvin temperature scale, k has the value 1.38×10^{-23} joules per kelvin. Energy is reported in joules (J): $1\ \text{J} = 1\ \text{kg m}^2\ \text{s}^{-2}$. We could think of 1 J as the energy of a 2 kg ball travelling at $1\ \text{m s}^{-1}$. Each pulse of the human heart expends an energy of about 1 J. The point to remember is that, because β is proportional to $1/T$, as the temperature goes up, β goes down, and vice versa.

There are several points worth making here. First, the huge importance of the Boltzmann distribution is that it reveals the molecular significance of temperature: *temperature is the parameter that tells us the most probable distribution of populations of molecules over the available states of a system at equilibrium.* When the temperature is high (β low), many states have significant populations; when the temperature is low (β high), only the states close to the lowest state have significant populations (Figure 4). Regardless of the actual values of the populations, they invariably follow an exponential distribution of the kind given by the Boltzmann expression. In terms of our balls-on-shelves analogy, low temperatures (high β) corresponds to our throwing the balls weakly at the shelves so that only the lowest are occupied. High temperatures (low β) corresponds to our throwing the balls vigorously at the shelves, so that even high shelves are populated significantly. Temperature, then, is just *a parameter that summarizes the relative populations of energy levels in a system at equilibrium.*

The second point is that β is a more natural parameter for expressing temperature than T itself. Thus, whereas later we shall see that absolute zero of temperature ($T = 0$) is unattainable in a finite number of steps, which may be puzzling, it is far less

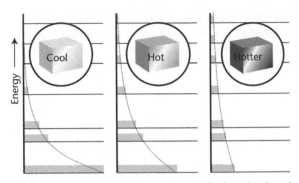

4. The Boltzmann distribution is an exponentially decaying function of the energy. As the temperature is increased, the populations migrate from lower energy levels to higher energy levels. At absolute zero, only the lowest state is occupied; at infinite temperature, all states are equally populated

surprising that an infinite value of β (the value of β when $T = 0$) is unattainable in a finite number of steps. However, although β is the more natural way of expressing temperatures, it is ill-suited to everyday use. Thus water freezes at 0°C (273 K), corresponding to $\beta = 2.65 \times 10^{20}$ J^{-1}, and boils at 100°C (373 K), corresponding to $\beta = 1.94 \times 10^{20}$ J^{-1}. These are not values that spring readily off the tongue. Nor are the values of β that typify a cool day (10°C, corresponding to 2.56×10^{20} J^{-1}) and a warmer one (20°C, corresponding to 2.47×10^{20} J^{-1}).

The third point is that the existence and value of the fundamental constant k is simply a consequence of our insisting on using a conventional scale of temperature rather than the truly fundamental scale based on β. The Fahrenheit, Celsius, and Kelvin scales are misguided: the reciprocal of temperature, essentially β, is more meaningful, more natural, as a measure of temperature. There is no hope, though, that it will ever be accepted, for history and the potency of simple numbers, like 0 and 100, and even 32 and 212, are too deeply embedded in our culture, and just too convenient for everyday use.

Although Boltzmann's constant k is commonly listed as a fundamental constant, it is actually only a recovery from a historical mistake. If Ludwig Boltzmann had done his work before Fahrenheit and Celsius had done theirs, then it would have been seen that β was the natural measure of temperature, and we might have become used to expressing temperatures in the units of inverse joules with warmer systems at low values of β and cooler systems at high values. However, conventions had become established, with warmer systems at higher temperatures than cooler systems, and k was introduced, through $k\beta = 1/T$, to align the natural scale of temperature based on β to the conventional and deeply ingrained one based on T. Thus, Boltzmann's constant is nothing but a conversion factor between a well-established conventional scale and the one that, with hindsight, society might have adopted. Had it adopted β as its measure of temperature, Boltzmann's constant would not have been necessary.

We shall end this section on a more positive note. We have established that the temperature, and specifically β, is a parameter that expresses the equilibrium distribution of the molecules of a system over their available energy states. One of the easiest systems to imagine in this connection is a perfect (or 'ideal') gas, in which we imagine the molecules as forming a chaotic swarm, some moving fast, others slow, travelling in straight lines until one molecule collides with another, rebounding in a different direction and with a different speed, and striking the walls in a storm of impacts and thereby giving rise to what we interpret as pressure. A gas is a chaotic assembly of molecules (indeed, the words 'gas' and 'chaos' stem from the same root), chaotic in spatial distribution and chaotic in the distribution of molecular speeds. Each speed corresponds to a certain kinetic energy, and so the Boltzmann distribution can be used to express, through the distribution of molecules over their possible translational energy states, their distribution of speeds, and to relate that distribution of speeds to the temperature. The resulting expression is called the *Maxwell–Boltzmann distribution* of speeds, for James Clerk

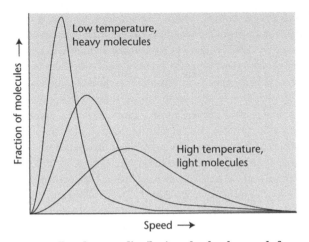

5. The Maxwell–Boltzmann distribution of molecular speeds for molecules of various mass and at different temperatures. Note that light molecules have higher average speeds than heavy molecules. The distribution has consequences for the composition of planetary atmospheres, as light molecules (such as hydrogen and helium) may be able to escape into space

Maxwell (1831–1879) first derived it in a slightly different way. When the calculation is carried through, it turns out that the average speed of the molecules increases as the square root of the absolute temperature. The average speed of molecules in the air on a warm day (25°C, 298 K) is greater by 4 per cent than their average speed on a cold day (0°C, 273 K). Thus, we can think of temperature as an indication of the average speeds of molecules in a gas, with high temperatures corresponding to high average speeds and low temperatures to lower average speeds (Figure 5).

A word of summary

A word or two of summary might be appropriate at this point. From the outside, from the viewpoint of an observer stationed, as always, in the surroundings, temperature is a property that reveals whether, when closed systems are in contact through diathermic boundaries, they will be in thermal equilibrium—their temperatures are the same—or whether there will be a consequent change of state—their temperatures are different—that will continue until the temperatures have equalized. From the inside, from the viewpoint of a microscopically eagle-eyed observer within the system, one able to discern the distribution of molecules over the available energy levels, the temperature is the single parameter that expresses those populations. As the temperature is increased, that observer will see the population extending up to higher energy states, and as it is lowered, the populations relax back to the states of lower energy. At any temperature, the relative population of a state varies exponentially with the energy of the state. That states of higher energy are progressively populated as the temperature is raised means that more and more molecules are moving (including rotating and vibrating) more vigorously, or the atoms trapped at their locations in a solid are vibrating more vigorously about their average positions. Turmoil and temperature go hand in hand.

Chapter 2
The first law

The conservation of energy

The first law of thermodynamics is generally thought to be the least demanding to grasp, for it is an extension of the *law of conservation of energy*, that energy can be neither created nor destroyed. That is, however much energy there was at the start of the universe, so there will be that amount at the end. But thermodynamics is a subtle subject, and the first law is much more interesting than this remark might suggest. Moreover, like the zeroth law, which provided an impetus for the introduction of the property 'temperature' and its clarification, the first law motivates the introduction and helps to clarify the meaning of the elusive concept of 'energy'.

We shall assume at the outset that we have no inkling that there is any such property, just as in the introduction to the zeroth law we did not pre-assume that there was anything we should call temperature, and then found that the concept was forced upon us as an implication of the law. All we shall assume is that the well-established concepts of mechanics and dynamics, like mass, weight, force, and work, are known. In particular, we shall base the whole of this presentation on an understanding of the notion of 'work'.

Work is motion against an opposing force. We do work when we raise a weight against the opposing force of gravity. The magnitude of the work we do depends on the mass of the object, the strength of the gravitational pull on it, and the height through which it is raised. You yourself might be the weight: you do work when you climb a ladder; the work you do is proportional to your weight and the height through which you climb. You also do work when cycling into the wind: the stronger the wind and the further you travel the greater the work you do. You do work when you stretch or compress a spring, and the amount of work you do depends on the strength of the spring and the distance through which it is stretched or compressed.

All work is equivalent to the raising of a weight. For instance, although we might think of stretching a spring, we could connect the stretched spring to a pulley and weight and see how far the weight is raised when the spring returns to its natural length. The magnitude of the work of raising a mass m (for instance, 50 kg) through a height h (for instance, 2.0 m) on the surface of the Earth is calculated from the formula mgh, where g is a constant known as the *acceleration of free fall*, which at sea level on Earth is close to 9.8 m s^{-2}. Raising a 50 kg weight through 2.0 m requires work of magnitude 980 kg m^2 s^{-2}. As we saw on p. 11, the awkward combination of units 'kilograms metre squared per second squared' is called the *joule* (symbol J). So, to raise our weight, we have to do 980 joules (980 J) of work.

Work is the primary foundation of thermodynamics and in particular of the first law. Any system has the capacity to do work. For instance, a compressed or extended spring can do work: as we have remarked, it can be used to bring about the raising of a weight. An electric battery has the capacity to do work, for it can be connected to an electric motor which in turn can be used to raise a weight. A lump of coal in an atmosphere of air can be used to do work by burning it as a fuel in some kind of engine. It is not an entirely obvious point, but when we drive an electric current through a heater, we are doing work on the heater, for the same current could be used to raise a weight by passing it through an electric motor rather than the heater. Why a heater is called a 'heater' and not a 'worker' will become clear once we have introduced the concept of heat. That concept hasn't appeared yet.

With work a primary concept in thermodynamics, we need a term to denote the capacity of a system to do work: that capacity we term *energy*. A fully stretched spring has a greater capacity to do work than the same spring only slightly stretched: the fully stretched spring has a greater energy than the slightly stretched spring. A litre of hot water has the capacity to do more work than the same litre of cold water: a litre of hot water has a greater energy than a litre of cold water. In this context, there is nothing mysterious about energy: it is just a measure of the capacity of a system to do work, and we know exactly what we mean by work.

Path independence

Now we extend these concepts from dynamics to thermodynamics. Suppose we have a system enclosed in adiabatic (thermally non-conducting) walls. We established the concept of 'adiabatic' in Chapter 1 by using the zeroth law, so we have not slipped in an undefined term. In practice, by 'adiabatic' we mean a thermally insulated container, like a well-insulated vacuum flask. We can monitor the temperature of the contents of the flask by using a thermometer, which is another concept introduced by the zeroth law, so we are still on steady ground. Now we do some experiments.

First, we churn the contents of the flask (that is, the system) with paddles driven by a falling weight, and note the change in temperature this churning brings about. Exactly this type of experiment was performed by J. P. Joule (1818–1889), one of the fathers of thermodynamics, in the years following 1843. We know how much work has been done by noting the heaviness of the weight and the distance through which it fell. Then we remove the insulation and let the system return to its original state. After replacing the insulation, we put a heater into the system and pass an electric current for a time that results in the same work being done on the heater as was done by the falling weight. We would have done other measurements to relate the current passing through a motor for various times and noting the height to which weights are raised, so we can interpret the combination of time and current as an amount of work performed. The conclusion we arrive at in this pair of experiments and in a multitude of others of a similar kind is that *the same amount of work, however it is performed, brings about the same change of state of the system.*

This conclusion is like climbing a mountain by a variety of different paths, each path corresponding to a different method of doing work. Provided we start at the same base camp and arrive at the same destination, we shall have climbed through the same height regardless of the path we took between them. That is, we can attach a number (the 'altitude') to every point on the mountain, and calculate the height we have climbed, regardless of the path, by taking the difference of the initial and final altitudes for our climb. Exactly the same applies to our system. The fact that the change of state is path-independent means that we can associate a number, which we shall call the *internal energy* (symbol U) with each state of the system. Then we can calculate the work needed to travel between any two states by taking the difference of the initial and final values of the internal energy, and write *work required* $= U(\text{final}) - U(\text{initial})$ (Figure 6).

The observation of the path-independence of the work required to go between two specified states in an adiabatic system (remember, at this stage the system is adiabatic) has motivated the recognition that there is a property of the system that is a measure of its capacity to do work. In thermodynamics, a property that depends

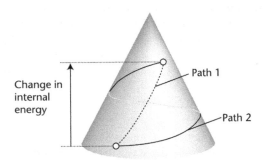

Change in
internal
energy

Path 1

Path 2

6. The observation that different ways of doing work on a system and thereby changing its state between fixed endpoints required the same amount of work is analogous to different paths on a mountain resulting in the same change of altitude leads to the recognition of the existence of a property called the internal energy

only on the current state of the system and is independent of how that state was prepared (like altitude in geography) is called a *state function*. Thus, our observations have motivated the introduction of the state function called internal energy. We might not understand the deep nature of internal energy at this stage, but nor did we understand the deep nature of the state function we called temperature when we first encountered it in the context of the zeroth law.

We have not yet arrived at the first law: this will take a little more work, both literally and figuratively. To move forward, let's continue with the same system but strip away the thermal insulation so that it is no longer adiabatic. Suppose we do our churning business again, starting from the same initial state and continuing until the system is in the same final state as before. We find that a different amount of work is needed to reach the final state.

Typically, we find that more work has to be done than in the adiabatic case. We are driven to conclude that the internal energy can change by an agency other than by doing work. One way of regarding this additional change is to interpret it as arising from the transfer of energy from the system into the surroundings due to the difference in temperature caused by the work that we do as we churn the contents. This transfer of energy as a result of a temperature difference is called *heat*.

The amount of energy that is transferred as heat into or out of the system can be measured very simply: we measure the work required to bring about a given change in the adiabatic system, and then the work required to bring about the same change of state in the diathermic system (the one with thermal insulation removed), and take the difference of the two values. That difference is the energy transferred as heat. A point to note is that the measurement of the rather elusive concept of 'heat' has been put on a purely mechanical foundation as the difference in the heights through which a weight falls to bring about a given change of state under two different conditions (Figure 7).

We are within a whisper of arriving at the first law. Suppose we have a closed system and use it to do some work or allow a release of energy as heat. Its internal energy falls. We then leave the

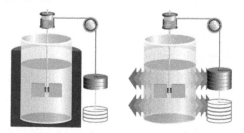

7. When a system is adiabatic (left), a given change of state is brought about by doing a certain amount of work. When the same system undergoes the same change of state in a non-adiabatic container (right), more work has to be done. The difference is equal to the energy lost as heat

system isolated from its surroundings for as long as we like, and later return to it. We invariably find that its capacity to do work—its internal energy—has not been restored to its original value. In other words,

the internal energy of an isolated system is constant.

That is the first law of thermodynamics, or at least one statement of it, for the law comes in many equivalent forms.

Another universal law of nature, this time of human nature, is that the prospect of wealth motivates deceit. Wealth—and untold benefits to humanity—would accrue to an untold extent if the first law were found to be false under certain conditions. It would be found to be false if work could be generated by an adiabatic, closed system without a diminution of its internal energy. In other words, if we could achieve *perpetual motion*, work produced without consumption of fuel. Despite enormous efforts, perpetual motion has never been achieved. There have been claims galore, of course, but all of them have involved a degree of deception. Patent offices are now closed to the consideration of all such machines, for the first law is regarded as unbreakable and reports of its transgression not worth the time or effort to pursue. There are certain instances in science, and certainly in technology, where a closed mind is probably justified.

Heat as energy in transition

We have a variety of cleaning-up exercises to do before we leave the first law. First, there is the use of the term 'heat'. In everyday language, heat is both a noun and a verb. Heat flows; we heat. In thermodynamics heat is not an entity or even a form of energy: *heat is a mode of transfer of energy*. It is not a form of energy, or a fluid of some kind, or anything of any kind. Heat is the transfer of energy by virtue of a temperature difference. Heat is the name of a process, not the name of an entity.

Everyday discourse would be stultified if we were to insist on the precise use of the word heat, for it is enormously convenient to speak of heat flowing from here to there, and to speak of heating an object. The first of these everyday usages was motivated by the view that heat is an actual fluid that flows between objects at different temperatures, and this powerful imagery is embedded indelibly in our language. Indeed, there are many aspects of the migration of energy down temperature gradients that are fruitfully treated mathematically by regarding heat as the flow of a massless ('imponderable') fluid. But that is essentially a coincidence, it is not an indicator that heat is actually a fluid any more than the spread of consumer choice in a population, which can also be treated by similar equations, is a tangible fluid.

What we should say, but it is usually too tedious actually to say it repeatedly, is that energy is transferred as heat (that is, as the result of a temperature difference). To heat, the verb, should for precision be replaced by circumlocutions such as 'we contrive a temperature difference such that energy flows through a diathermic wall in a desired direction'. Life, though, is too short, and it is expedient, except when we want to be really precise, to adopt the casual easiness of everyday language, and we shall cross our fingers and do so, but do bear in mind how that shorthand should be interpreted.

Heat and work: a molecular view

There has probably been detected a slipperiness in the preceding remarks, for although we have warned against regarding heat as a fluid, there is still the whiff of fluidity about our use of the term energy. It looks as though we have simply pushed back to a deeper layer the notion of fluid. This apparent deceit, though, is resolved by identifying the molecular natures of heat and work. As usual, digging into the underworld of phenomena illuminates them. In thermodynamics, we always distinguish between the modes of

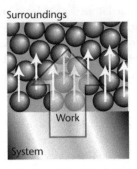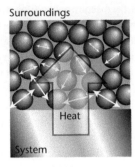

8. The molecular distinction between the transfer of energy as work (left) and heat (right). Doing work results in the uniform motion of atoms in the surroundings; heating stimulates their disorderly motion

transfer of energy by observations in the surroundings: the system is blind to the processes by which it is provided with or loses energy. We can think of a system as like a bank: money can be paid in or withdrawn in either of two currencies, but once inside there is no distinction between the type of funds in which its reserves are stored.

First, the molecular nature of work. We have seen that at an observational level, doing work is equivalent to the raising of a weight. From a molecular viewpoint, the raising of a weight corresponds to all its atoms moving in the same direction. Thus, when a block of iron is raised, all the iron atoms move upwards uniformly. When the block is lowered—and does work on the system, like compressing a spring or a gas, and increases its internal energy—all its atoms move downwards uniformly. *Work is the transfer of energy that makes use of the uniform motion of atoms in the surroundings* (Figure 8).

Now, the molecular nature of heat. We saw in Chapter 1 that the temperature is a parameter that tells us the relative numbers of

24

atoms in the allowed energy states, with the higher energy states progressively more populated as the temperature is increased. In more pictorial terms, a block of iron at high temperature consists of atoms that are oscillating vigorously around their average positions. At low temperatures, the atoms continue to oscillate, but with less vigour. If a hot block of iron is put in contact with a cooler block, the vigorously oscillating atoms at the edge of the hot block jostle the less vigorously oscillating atoms at the edge of the cool block into more vigorous motion, and they pass on their energy by jostling their neighbours. There is no net motion of either block, but energy is transferred from the hotter to the cooler block by this random jostling where the two blocks are in contact. That is, *heat is the transfer of energy that makes use of the random motion of atoms in the surroundings* (Figure 8).

Once the energy is inside the system, either by making use of the uniform motion of atoms in the surroundings (a falling weight) or of randomly oscillating atoms (a hotter object, such as a flame), there is no memory of how it was transferred. Once inside, the energy is stored as the kinetic energy (the energy due to motion) and the potential energy (the energy due to position) of the constituent atoms, and that energy can be withdrawn either as heat or as work. The distinction between work and heat is made in the surroundings: the system has no memory of the mode of transfer nor is it concerned about how its store of energy will be used.

This blindness to the mode of transfer needs a little further explanation. Thus, if a gas in an adiabatic container is compressed by a falling weight, the incoming piston acts like a bat in a microscopic game of table-tennis. When a molecule strikes the piston, it is accelerated. However, as it flies back into the gas it undergoes collisions with the other molecules in the system, and as a result its enhanced kinetic energy is quickly dispersed over them and its direction of motion is randomized. When the same sample of gas is heated, the random jostling of the atoms in the surroundings stimulates the gas molecules into more vigorous motion, and the acceleration of the molecules at the thermally conducting walls is quickly distributed over the entire sample. The result within the system is the same.

We can now return to the faintly enigmatic remark made earlier that an electric heater is better regarded as an electric worker. The electric current that is passed through the coil of wire within the heater is a uniform flow of electrons. The electrons of that current collide with the atoms of the wire and cause them to wobble around their mean positions. That is, the energy—and the temperature—of the coil of wire is raised by doing work on it. However, the coil of wire is in thermal contact with the contents of the system, and the vigorous motion of the atoms of the wire jostle the atoms of the system; that is, the filament heats the system. So, although we do work on the heater itself, that work is translated into heating the system: worker has become heater.

A final point is that the molecular interpretation of heat and work elucidates one aspect of the rise of civilization. Fire preceded the harnessing of fuels to achieve work. The heat of fire—the tumbling out of energy as the chaotic motion of atoms—is easy to contrive for the tumbling is unconstrained. Work is energy tamed, and requires greater sophistication to contrive. Thus, humanity stumbled easily on to fire but needed millennia to arrive at the sophistication of the steam engine, the internal combustion engine, and the jet engine.

Introducing reversibility

The originators of thermodynamics were subtle people, and quickly realized that they had to be careful when specifying how a process is carried out. Although the technicality we shall describe now has little immediate relevance to the first law at the level of our discussion, it will prove to be of vital significance when we turn to the second law.

I alluded in Chapter 1 to science's hijacking of familiar words and adding a new precision to their meaning. In the current context we need to consider the word 'reversible'. In everyday language, a reversible process is one that can be reversed. Thus, the rolling of a wheel can be reversed, so in principle a journey can be traversed in reverse. The compression of a gas can be reversed by pulling out the piston that effected the compression. In thermodynamics 'reversible' means something rather more refined: a *reversible process* in thermodynamics is one that is reversed by an *infinitesimal* modification of the conditions in the surroundings.

The key word is infinitesimal. If we think of a gas in a system at a certain pressure, with the piston moving out against a lower external pressure, an infinitesimal change in the external pressure will not reverse the motion of the piston. The expansion is reversible in the colloquial sense but not in the thermodynamic sense. If a block of iron (the system) at 20°C is immersed in a water bath at 40°C, energy will flow as heat from the bath into the block, and an infinitesimal change in the temperature of the water will have no effect on the direction of flow. The transfer of energy as heat is not reversible in the thermodynamic sense in this instance. However, now consider the case in which the external pressure matches the pressure of the gas in the system exactly. As we saw in Chapter 1, we say that the system and its surroundings are in mechanical equilibrium. Now increase the external pressure infinitesimally: the piston moves in a little. Now reduce the external pressure infinitesimally: the piston moves out a little. We see that the direction of motion of the piston is changed by an infinitesimal change in a property, in this case the pressure, of the surroundings. The expansion is reversible in the thermodynamic sense. Likewise, consider a system at the same temperature as the surroundings. In this case the system and its surroundings are in thermal equilibrium. If we reduce the temperature of the surroundings infinitesimally, energy flows out of the system as heat. If we increase the temperature of the surroundings infinitesimally, energy flows into the system as heat. The transfer of energy as heat is reversible in the thermodynamic sense in this instance.

The greatest amount of work is done if the expansion of a gas is reversible at every stage. Thus, we match the external pressure to the pressure of the gas in the system and then reduce the external pressure infinitesimally: the piston moves out a little. The pressure of the gas falls a little because it now occupies a greater volume. Then we reduce the external pressure infinitesimally, the piston moves out a little more and the pressure of the gas decreases a little. This process of effectively matching the external pressure to the falling pressure of the gas continues until the piston has moved out a desired amount and, through its coupling to a weight, has done a certain amount of work. No greater work can be done, because if at any stage the external pressure is increased even infinitesimally, then the piston will move in rather than out. That is, by ensuring that at every stage the expansion is reversible in the thermodynamic sense, the system does maximum work. This conclusion is general: *reversible changes achieve maximum work*. We shall draw on this generalization in the following chapters.

Introducing enthalpy

Thermodynamicists are also subtle in their discussion of the quantity of heat that can be extracted from a system, such as when a fuel burns. We can appreciate the problem as follows. Suppose we burn a certain amount of hydrocarbon fuel in a container fitted with a movable piston. As the fuel burns it produces carbon dioxide and water vapour, which occupy more space than the original fuel and oxygen, and as a result the piston is driven out to accommodate the products. This expansion requires work. That is, when a fuel burns in a container that is free to expand, some of the energy released in the combustion is used to do work. If the combustion takes place in a container with rigid walls, the combustion releases the same amount of energy, but none of it is used to do work because no expansion can occur. In other words, more energy is available as heat in the latter case than in the former. To calculate the heat that can be produced in the former case, we have to account for the energy that is used to make room for the carbon dioxide and water vapour and subtract that from the total change in energy. This is true even if there is no physical piston—if the fuel burns in a dish—because, although we cannot see it so readily, the gaseous products must still make room for themselves.

Thermodynamicists have developed a clever way of taking into account the energy used to do work when any change, and particularly the combustion of a fuel, occurs, without having to calculate the work explicitly in each case. To do so, they switch attention from the internal energy of a system, its total energy content, to a closely related quantity, the *enthalpy* (symbol H). The name comes from the Greek words for 'heat inside', and although, as we have stressed, there is no such thing as 'heat' (it is a process of transfer, not a thing), for the circumspect the name is well chosen, as we shall see. The formal relation of enthalpy, H, to internal energy, U, is easily written down as $H = U + pV$, where p is the pressure of the system and V is its volume. From this relation it follows that the enthalpy of a litre of water open to the atmosphere is only 100 J greater than its internal energy, but it is much more important to understand its significance than to note small differences in numerical values.

It turns out that the energy released as heat by a system free to expand or contract as a process occurs, as distinct from the total energy released in the same process, is exactly equal to the change in enthalpy of the system. That is, as if by magic—but actually by mathematics—the leakage of energy from a system as work is automatically taken into account by focusing on the change in enthalpy. In other words, the enthalpy is the basis of a kind of accounting trick, which keeps track invisibly of the work that is done by the system, and reveals the amount of energy that is released only as heat, provided the system is free to expand in an atmosphere that exerts a constant pressure on the system.

It follows that if we are interested in the heat that can be generated by the combustion of a fuel in an open container, such as a furnace, then we use tables of enthalpies to calculate the change in enthalpy that accompanies the combustion. This change is written ΔH, where the Greek uppercase delta is used throughout thermodynamics to denote a change in a quantity. Then we identify that change with the heat generated by the system. As an actual example, the change of enthalpy that accompanies the combustion of a litre of gasoline is about 33 megajoules (1 megajoule, written 1 MJ, is 1 million joules). Therefore we know without any further calculation that burning a litre of gasoline in an open container will provide 33 MJ of heat. A deeper analysis of the process shows that in the same combustion, the system has to do about 130 kJ (where 1 kilojoule, written 1 kJ, is one thousand joules) of work to make room for the gases that are generated, but that energy is not available to us as heat.

We could extract that extra 130 kJ, which is enough to heat about half a litre of water from room temperature to its boiling point, if we prevent the gases from expanding so that all the energy released in the combustion is liberated as heat. One way to achieve that, and to obtain all the energy as heat, would be to arrange for the combustion to take place in a closed container with rigid walls, in which case it would be unable to expand and hence would be unable to lose any energy as work. In practice, it is technologically much simpler to use furnaces that are open to the atmosphere, and in practice the difference between the two cases is too small to be worth the effort. However, in formal thermodynamics, which is a precise and logical subject, it is essential to do all the energy accounting accurately and systematically. In formal thermodynamics the differences between changes in internal energy and enthalpy must always be borne in mind.

The vaporization of a liquid requires an input of energy because its molecules must be separated from one another. This energy is commonly supplied in the form of heat—that is, by making use of a temperature difference between the liquid and its surroundings. In former times, the extra energy of the vapour was termed the 'latent heat', because it was released when the vapour re-condensed to a liquid and was in some sense 'latent' in the vapour. The scalding effect of steam is an illustration. In modern thermodynamic terms, the supply of energy as heat is identified with the change in enthalpy of the liquid, and the term 'latent heat' has been replaced by *enthalpy of vaporization*. The enthalpy of vaporization of 1 g of water is close to 2 kJ. The condensation of 1 g of steam therefore releases 2 kJ of heat, which may be enough to destroy the proteins of our skin where it comes in contact. There is a corresponding term for the heat required to melt a solid: the 'enthalpy of fusion'. Gram-for-gram, the enthalpy of fusion is much less than the enthalpy of vaporization, and we do not get scalded by touching water that is freezing to ice.

Heat capacity

We saw in Chapter 1 in the context of the zeroth law that 'temperature' is a parameter that tells us the occupation of the available energy levels of the system. Our task now is to see how this zeroth-law property relates to the first-law property of internal energy and the derived heat-accounting property of enthalpy.

As the temperature of a system is raised and the Boltzmann distribution acquires a longer tail, populations migrate from states of lower energy to states of higher energy. Consequently, the *average* energy rises, for its value takes into account the energies of the available states and the numbers of molecules that occupy each one. In other words, as the temperature is raised, so the internal energy rises. The enthalpy rises too, but we don't need to focus on that separately as it more or less tracks the changes in internal energy.

The slope of a graph of the value of the internal energy plotted against temperature is called the *heat capacity* of the system (symbol C). To be almost precise, the heat capacity is defined as C = (heat supplied)/(resulting temperature rise). The supply of 1 J of energy as heat to 1 g of water results in an increase in temperature of about $0.2°C$. Substances with a high heat capacity (water is an example) require a larger amount of heat to bring about a given rise in temperature than those with a small heat capacity (air is an example). In formal thermodynamics, the conditions under which heating takes place must be specified. For instance, if the heating takes place under conditions of constant pressure with the sample free to expand, then some of the energy supplied as heat goes into expanding the sample and therefore to doing work. Less energy remains in the sample, so its temperature rises less than when it is constrained to have a constant volume, and therefore we report that its heat capacity is higher. The difference between heat capacities of a system at constant volume and at constant pressure is of most practical significance for gases, which undergo large changes in volume as they are heated in vessels that are able to expand.

Heat capacities vary with temperature. An important experimental observation that will play an important role in the following chapter is that the heat capacity of every substance falls to zero when the temperature is reduced towards absolute zero ($T = 0$). A very small heat capacity implies that even a tiny transfer of heat to a system results in a significant rise in temperature, which is one of the problems associated with achieving very low temperatures when even a small leakage of heat into a sample can have a serious effect on the temperature (see Chapter 5).

We can get insight into the molecular origin of heat capacity by thinking—as always—about the distribution of molecules over the available states. There is a deep theorem of physics called the *fluctuation–dissipation theorem*, which implies that the ability of a system to dissipate (essentially, absorb) energy is proportional to the magnitudes of the fluctuations about its mean value in a corresponding property. Heat capacity is a dissipation term: it is a measure of the ability of a substance to absorb the energy supplied to it as heat. The corresponding fluctuation term is the spread of populations over the energy states of the system. When all the molecules of a system are in a single state, there is no spread of populations and the 'fluctuation' in population is zero; correspondingly the heat capacity of the system is zero. As we saw in Chapter 1, at $T = 0$ only the lowest state of the system is occupied, so we can conclude from the fluctuation–dissipation theorem that the heat capacity will be zero too, as is observed. At higher temperatures, the populations are spread over a range of states and hence the heat capacity is non-zero, as is observed.

In most cases, the spread of populations increases with increasing temperature, so the heat capacity typically increases with rising temperature, as is observed. However, the relationship is a little more complex than that because it turns out that the role of the spread of populations decreases as the temperature rises, so although that spread increases, the heat capacity does not increase as fast. In some cases, the increasing spread is balanced exactly by the decrease in the proportionality constant that relates the spread to the heat capacity, and the heat capacity settles into a constant value. This is the case for the contribution of all the basic modes of motion: translation (motion through space), rotation, and vibration of molecules, all of which settle into a constant value.

To understand the actual values of the heat capacity of a substance and the rise in internal energy as the temperature is raised we first need to understand how the energy levels of a substance depend on its structure. Broadly speaking, the energy levels lie close together when the atoms are heavy. Moreover, translational energy levels are so close together as to form a near continuum, the rotational levels of molecules in gases are further apart, and vibrational energy levels—those associated with the oscillations of atoms within molecules—are widely separated. Thus, as a gaseous sample is heated, the molecules are readily excited into higher translational states (in English: they move faster) and, in all practical cases, they quickly spread over many rotational states (in English: they rotate faster). In each case the average energy of the molecules, and hence the internal energy of the system, increases as the temperature is raised.

The molecules of solids are free neither to move through space nor to rotate. However, they can oscillate around their average positions, and take up energy that way. These collective wobblings of the entire solid have much lower frequencies than the oscillations of atoms within molecles and so they can be excited much more readily. As energy is supplied to a solid sample, these oscillations are excited, the populations of the higher energy states increase as the Boltzmann distribution reaches to higher levels, and we report that the temperature of the solid has risen. Similar remarks apply to liquids, in which molecular motion is less constrained than in solids. Water has a very high heat capacity, which means that to raise its temperature takes a lot of energy. Conversely, hot water stores a lot of energy, which is why it is such a good medium for central heating systems (as well as being cheap), and why the oceans are slow to heat and slow to cool, with important implications for our climate.

As we have remarked, the internal energy is simply the total energy of the system, the sum of the energies of all the molecules and their interactions. It is much harder to give a molecular interpretation of enthalpy because it is a property contrived to do the bookkeeping of expansion work and is not as fundamental a property as internal energy. For the purposes of this account, it is best to think of the enthalpy as a measure of the total energy, but to bear in mind that that is not exactly true. In short, as the temperature of a system is raised its molecules occupy higher and higher energy levels and as a result their mean energy, the internal energy, and the enthalpy all increase. Precise fundamental molecular interpretations can be given only of the fundamental properties of a system, its temperature, its internal energy, and—as we shall see in the next chapter—the entropy. They cannot be given for 'accounting' properties, properties that have simply been contrived to make calculations easier.

Energy and the uniformity of time

The first law is essentially based on the conservation of energy, the fact that energy can be neither created nor destroyed. Conservation laws—laws that state that a certain property does not change—have a very deep origin, which is one reason why scientists, and thermodynamicists in particular, get so excited when nothing happens. There is a celebrated theorem, *Noether's theorem*, proposed by the German mathematician Emmy Noether (1882–1935), which states that to every conservation law there corresponds a symmetry. Thus, conservation laws are based on various aspects of the shape of the universe we inhabit. In the particular case of the conservation of energy, the symmetry is that of the shape of time. Energy is conserved because time is uniform:

time flows steadily, it does not bunch up and run faster then spread out and run slowly. Time is a uniformly structured coordinate. If time were to bunch up and spread out, energy would not be conserved. Thus, the first law of thermodynamics is based on a very deep aspect of our universe and the early thermodynamicists were unwittingly probing its shape.

Chapter 3
The second law

The increase in entropy

When I gave lectures on thermodynamics to an undergraduate chemistry audience I often began by saying that no other scientific law has contributed more to the liberation of the human spirit than the second law of thermodynamics. I hope that you will see in the course of this chapter why I take that view, and perhaps go so far as to agree with me.

The second law has a reputation for being recondite, notoriously difficult, and a litmus test of scientific literacy. Indeed, the novelist and former chemist C. P. Snow is famous for having asserted in his *The Two Cultures* that not knowing the second law of thermodynamics is equivalent to never having read a work by Shakespeare. I actually have serious doubts about whether Snow understood the law himself, but I concur with his sentiments. The second law is of central importance in the whole of science, and hence in our rational understanding of the universe, because it provides a foundation for understanding why *any* change occurs. Thus, not only is it a basis for understanding why engines run and chemical reactions occur, but it is also a foundation for understanding those most exquisite consequences of chemical reactions, the acts of literary, artistic, and musical creativity that enhance our culture.

As we have seen for the zeroth and first laws, the formulation and interpretation of a law of thermodynamics leads us to introduce a thermodynamic property of the system: the temperature, T, springs from the zeroth law and the internal energy, U, from the first law. Likewise, the second law implies the existence of another thermodynamic property, the *entropy* (symbol S). To fix our ideas in the concrete at an early stage it will be helpful throughout this account to bear in mind that whereas U is a measure of the *quantity* of energy that a system possesses, S is a measure of the *quality* of that energy: low entropy means high quality; high entropy means low quality. We shall elaborate this interpretation and show its consequences in the rest of the chapter. At the end of it, with the existence and properties of T, U, and S established, we shall have completed the foundations of classical thermodynamics in the sense that the whole of the subject is based on these three properties.

A final point in this connection, one that will pervade this chapter, is that power in science springs from abstraction. Thus, although a feature of nature may be established by close observation of a concrete system, the scope of its application is extended enormously by expressing the observation in abstract terms. Indeed, we shall see in this chapter that although the second law was established by observations on the lumbering cast-iron reality of a steam engine, when expressed in abstract terms it applies to all change. To put it another way, a steam engine encapsulates the nature of change whatever the concrete (or cast-iron) realization of that change. All our actions, from digestion to artistic creation, are at heart captured by the essence of the operation of a steam engine.

Heat engines

A steam engine, in its actual but not abstract form, is an iron fabrication, with boiler, valves, pipes, and pistons. The essence of a steam engine, though, is somewhat simpler: it consists of a hot (that is, high temperature) source of energy, a device—a piston or turbine—for converting heat into work, and a cold sink, a place for discarding any unused energy as heat. The last item, the cold sink, is not always readily discernible, for it might just be the immediate environment of the engine, not something specifically designed.

In the early nineteenth century, the French were anxiously observing from across the Channel England's industrialization and becoming envious of her increasing efficiency at using her abundant supplies of coal to pump water from her mines and drive her emerging factories. A young French engineer, Sadi Carnot (1796–1832), sought to contribute to his country's economic and military might by analysing the constraints on the efficiency of a steam engine. Popular wisdom at the time looked for greater efficiency in choosing a different working substance—air, perhaps, rather than steam—or striving to work at dangerously higher pressures. Carnot took the then accepted view that heat was a kind of imponderable fluid that, as it flowed from hot to cold, was able to do work, just as water flowing down a gradient can turn a water mill. Although his model was wrong, Carnot was able to arrive at a correct and astonishing result: that the efficiency of a perfect steam engine is independent of the working substance and depends only on the temperatures at which heat is supplied from the hot source and discarded into the cold sink.

The 'efficiency' of a steam engine—in general, a heat engine—is defined as the ratio of the work it produces to the heat it absorbs. Thus, if all the heat is converted into work, with none discarded, the efficiency is 1. If only half the supplied energy is converted into work, with the remaining half discarded into the surroundings, then the efficiency is 0.5 (which would commonly be reported as a percentage, 50 per cent). Carnot was able to derive the following expression for the maximum efficiency of an engine working between the absolute temperatures T_{source} and T_{sink}:

$$\text{Efficiency} = 1 - \frac{T_{sink}}{T_{source}}$$

This remarkably simple formula applies to any thermodynamically perfect heat engine regardless of its physical design. It gives the maximum theoretical efficiency, and no tinkering with a sophisticated design can increase the efficiency of an actual heat engine beyond this limit.

For instance, suppose a power station provided superheated steam to its turbines at 300°C (corresponding to 573 K) and allows the waste heat to spread into the surroundings at 20°C (293 K), the maximum efficiency is 0.46, so only 46 per cent of the heat supplied by the burning fuel can be converted into electricity, and no amount of sophisticated engineering design can improve on that figure given the two temperatures. The only way to improve the conversion efficiency would be to lower the temperature of the surroundings, which in practice is not possible in a commercial installation, or to use steam at a higher temperature. To achieve 100 per cent efficiency, the surroundings would have to be at absolute zero ($T_{sink} = 0$) or the steam would have to be infinitely hot ($T_{source} = \infty$), neither of which is a practical proposition.

Carnot's analysis established a very deep property of heat engines, but its conclusion was so alien to the engineering prejudices of the time that it had little impact. Such is often the fate of rational thought within society, sent as it may be to purgatory for a spell. Later in the century, and largely oblivious of Carnot's work, interest in heat was rekindled and two intellectual giants strode on to the stage and considered the problem of change, and in particular the conversion of heat into work, from a new perspective.

The first giant, William Thomson, later Lord Kelvin (1824–1907), reflected on the essential structure of heat engines. Whereas lesser

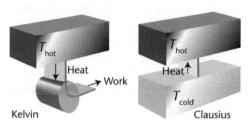

9. The Kelvin (left) and Clausius (right) observations are, respectively, that a cold sink is essential to the operation of a heat engine and that heat does not flow spontaneously from a cooler to a hotter body

minds might view the heat source as the crucial component, or perhaps the vigorously reciprocating piston, Kelvin—as we shall slightly anachronistically call him—saw otherwise: he identified the invisible as indispensible, seeing that the cold sink—often just the undesigned surroundings—is essential. Kelvin realized that to take away the surroundings would stop the heat engine in its tracks. To be more precise, the *Kelvin statement* of the second law of thermodynamics is as follows (Figure 9):

> no cyclic process is possible in which heat is taken from a hot source and converted completely into work.

In other words, Nature exerts a tax on the conversion of heat into work, some of the energy supplied by the hot source must be paid into the surroundings as heat. There must be a cold sink, even though we might find it hard to identify and it is not always an engineered part of the design. The cooling towers of a generating station are, in this sense, far more important to its operation than the complex turbines or the expensive nuclear reactor that seems to drive them.

The second giant was Rudolph Clausius (1822–1888), working in Berlin. He reflected on a simpler process, the flow of heat between bodies at different temperatures. He recognized the everyday phenomenon that energy flows as heat spontaneously from a body at a high temperature to one at a lower temperature. The word 'spontaneous' is another of those common words that has been captured by science and dressed in a more precise meaning. In thermodynamics *spontaneous* means not needing to be driven by doing work of some kind. Broadly speaking, 'spontaneous' is a synonym of 'natural'. Unlike in everyday language, spontaneous in thermodynamics has no connotation of speed: it does not mean fast. Spontaneous in thermodynamics refers to the *tendency* for a change to occur. Although some spontaneous processes are fast (the free expansion of a gas for instance) some may be immeasurably slow (the conversion of diamond into graphite, for instance). Spontaneity is a thermodynamic term that refers to a tendency, not necessarily to its actualization. Thermodynamics is silent on rates. For Clausius, there is a tendency for energy to flow as heat from high temperature to low, but the spontaneity of that process might be thwarted if an insulator lies in the way.

Clausius went on to realize that the reverse process, the transfer of heat from a cold system to a hotter one—that is, from a system at a low temperature to one at a higher temperature—is not spontaneous. He thereby recognized an asymmetry in Nature: although energy has a tendency to migrate as heat from hot to cold, the reverse migration is not spontaneous. This somewhat obvious statement he formulated into what is now known as the *Clausius statement* of the second law of thermodynamics (Figure 9):

> heat does not pass from a body at low temperature to one at high temperature without an accompanying change elsewhere.

In other words, heat can be transferred in the 'wrong' (non-spontaneous) direction, but to achieve that transfer work must be done. That is an everyday observation: we can cool objects in a refrigerator, which involves transferring heat from them and depositing it in the warmer surroundings, but to do so, we have to do work—the refrigerator must be driven by connecting it to a power supply, and the ultimate change elsewhere in the surroundings that drives the refrigeration is the combustion of fuel in a power station that may be far away.

The Kelvin and the Clausius statements are both summaries of observations. No one has ever built a working heat engine without a cold sink, although they might not have realized one was present. Nor have they observed a cool object spontaneously becoming hotter than its surroundings. As such, their statements are indeed laws of Nature in the sense that I am using the term as a summary of exhaustive observations. But are there two second laws? Why is Kelvin's, for instance, not called the second law and Clausius's the third?

The answer is that the two statements are logically equivalent. That is, Kelvin's statement implies Clausius's and Clausius's statement implies Kelvin's. I shall now demonstrate both sides of this equivalence.

First, imagine coupling two engines together (Figure 10). The two engines share the same hot source. Engine A has no cold sink, but engine B does. We use engine A to drive engine B. We run engine A, and for the moment presume, contrary to Kelvin's statement, that all the heat that A extracts from the hot source is converted into work. That work is used to drive the transfer of heat from the cold sink of engine B into the shared hot sink. The net effect is the restoration of the energy to the hot sink in addition to whatever engine B transferred out of its cold sink. That is, heat has been transferred from cold to hot with no change elsewhere, which is contrary to Clausius's statement. Therefore, if Kelvin's statement were ever found to be false, then Clausius's statement would be falsified too.

Now consider the implication of a failure of Clausius's statement. We build an engine with a hot source and a cold sink, and run the engine to produce work. In the process we discard some heat into

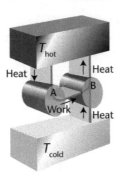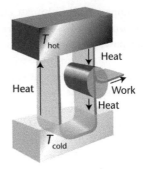

10. The equivalence of the Kelvin and Clausius statements. The diagram on the left depicts the fact that the failure of the Kelvin statement implies the failure of the Clausius statement. The diagram on the right depicts the fact that the failure of the Clausius statement implies the failure of the Kelvin statement

the cold sink. However, as a cunning part of the design we have also arranged for exactly the same amount of heat that we discarded into the cold sink to return spontaneously, contrary to Clausius's statement, to the hot source. Now the net effect of this arrangement is the conversion of heat into work with no other change elsewhere, for there is no net change in the cold sink, which is contrary to Kelvin's statement. Thus, if Clausius's statement were ever found to be false, then Kelvin's statement would be falsified too.

We have seen that the falsification of each statement of the second law implies the other, so logically the two statements are equivalent, and we can treat either as an equivalent phenomenological (observation-based) statement of the second law of thermodynamics.

The definition of absolute temperature

An interesting side issue is that the discussion so far enables us to set up a temperature scale that is based purely on mechanical observations, with the notion of a thermometer built solely from weights, ropes, and pulleys. You will recall that the zeroth law implied the existence of a property that we call the temperature, but apart from the arbitrary scales of Celsius and Fahrenheit, and a mention of the existence of a more fundamental thermodynamic scale, the definition was left hanging. Kelvin realized that he could define a temperature scale in terms of work by using Carnot's expression for the efficiency of a heat engine.

We shall denote the efficiency, the work done divided by heat absorbed, of a perfect heat engine by ε (the Greek letter epsilon). The work done by the engine can be measured by observing the height through which a known weight is raised, as we have already seen in the discussion of the first law. The heat absorbed by the engine can also, in principle at least, be measured by measuring the fall in a weight. Thus, as we saw in Chapter 2, the transfer of energy as heat can be measured by observing how much work must be done to achieve a given change of state in an adiabatic container, then measuring the work that must be done to achieve the same change in a diathermic container, and identifying the difference of the two amounts of work as the heat transaction in the second process. Thus, in principle, the efficiency of a heat engine can be measured solely by observing the rise or fall of a weight in a series of experiments.

Next, according to Carnot's expression, which in terms of ε is $\varepsilon = 1 - T_{sink}/T_{source}$, we can write $T_{sink}/T_{source} = 1 - \varepsilon$, or $T_{sink} = (1 - \varepsilon)T_{source}$. Therefore, to measure the temperature of the cold sink we simply use our weights to measure the efficiency of an engine that uses it. Thus, if we find $\varepsilon = 0.240$, then the temperature of the cold sink must be $0.760\,T_{source}$.

This still leaves T_{source} unspecified. We can choose a highly reproducible system, one more reliable than Fahrenheit's armpit, and define its temperature as having a certain value, and use that standard system as the hot source in the engine. In modern work, a system in which pure liquid water is simultaneously in equilibrium with both its vapour and ice, the so called *triple point* of water, is defined as having a temperature of exactly 273.16 K. The triple point is a fixed property of water: it is unaffected by any changes in the external conditions, such as the pressure, so it is highly reproducible. Therefore, in our example, if we measured by a series of observations on falling weights the efficiency of a heat engine that had a hot source at the temperature of the triple point of water, and found $\varepsilon = 0.240$, we would be able to infer that the temperature of the cold sink was 0.760×273.16 K = 208 K (corresponding to $-65°C$). The choice of the triple point of water for defining the Kelvin scale is entirely arbitrary, but it has the advantage that anyone in the galaxy can replicate the scale without any ambiguity, because water has the same properties everywhere without our having to adjust any parameters.

The everyday Celsius scale is currently defined in terms of the more fundamental thermodynamic scale by subtracting exactly 273.15 K from the Kelvin temperature. Thus, at atmospheric pressure, water is found to freeze at 273 K (to be precise, at about 0.01 K below the triple point, at close to 273.15 K), which corresponds to 0°C. Water is found to boil at 373 K, corresponding to close to 100°C. However, these two temperatures are no longer definitions, as they were when Anders Celsius proposed his scale in 1742, and must be determined experimentally. Their precise values are still open to discussion, but reliable values appear to be 273.152 518 K (+0.002 518°C) for the normal freezing point of water and 373.124 K (99.974°C) for its normal boiling point.

A final point is that the thermodynamic temperature is also occasionally called the 'perfect gas temperature'. The latter name comes from expressing temperature in terms of the properties of a perfect gas, a hypothetical gas in which there are no interactions between the molecules. That definition turns out to be identical to the thermodynamic temperature.

Introducing entropy

It is inelegant, but of practical utility, to have alternative statements of the second law. Our challenge is to find a single succinct statement that encapsulates them both. To do so, we follow Clausius and introduce a new thermodynamic function, the *entropy, S*. The etymology of the name, from the Greek words for 'in turning' is not particularly helpful; the choice of the letter S, which does from its shape suggest an 'in turning' appears, however, to be arbitrary, being a letter not used at the time for other thermodynamic properties, conveniently towards the end of the alphabet, and an unused neighbour of P, Q, R, T, U, V and W, all of which had already been ascribed other duties.

For mathematically cogent reasons that need not detain us here, Clausius defined a change in entropy of a system as the result of dividing the energy transferred as heat by the (absolute, thermodynamic) temperature at which the transfer took place:

$$\text{Change in entropy} = \frac{\text{heat supplied reversibly}}{\text{temperature}}$$

I have slipped in the qualification 'reversibly', because it is important, as we shall see, that the transfer of heat be imagined as carried out with only an infinitesimal difference in temperature between the system and its surroundings. In short, it is important not to stir up any turbulent regions of thermal motion.

We mentioned at the start of the chapter that entropy will turn out to be a measure of the 'quality' of the stored energy. As this chapter unfolds we shall see what 'quality' means. For our initial encounter with the concept, we shall identify entropy with disorder: if matter and energy are distributed in a disordered way, as in a gas, then the entropy is high; if the energy and matter are stored in an ordered manner, as in a crystal, then the entropy is low. With disorder in mind, we shall explore the implications of Clausius's expression and verify that it is plausible in capturing the entropy as a measure of the disorder in a system.

The analogy I have used elsewhere to help make plausible Clausius's definition of the change in entropy is that of sneezing in a busy street or in a quiet library. A quiet library is the metaphor for a system at low temperature, with little disorderly thermal motion. A sneeze corresponds to the transfer of energy as heat. In a quiet library a sudden sneeze is highly disruptive: there is a big increase in disorder, a large increase in entropy. On the other hand, a busy street is a metaphor for a system at high temperature, with a lot of thermal motion. Now the same sneeze will introduce relatively little additional disorder: there is only a small increase in entropy. Thus, in each case it is plausible that a change in entropy should be inversely proportional to some power of the temperature (the first power, T itself, as it happens; not T^2 or anything more complicated), with the greater change in entropy occurring the lower the temperature. In each case, the additional disorder is proportional to the magnitude of the sneeze (the quantity of energy transferred as heat) or some power of that quantity (the first power, as it happens). Thus, Clausius's expression conforms to this simple analogy, and we should bear the analogy in mind for the rest of the chapter as we see how to apply the concept of entropy and enrich our interpretation of it.

A change in entropy is the ratio of energy (in joules) transferred as heat to or from a system to the temperature (in kelvins) at which it is transferred, so its units are joules per kelvin ($J\ K^{-1}$). For instance, suppose we immerse a 1 kW heater in a tank of water at 20°C (293 K), and run the heater for 10 s, we increase the entropy of the water by 34 $J\ K^{-1}$. If 100 J of energy leaves a flask of water at 20°C, its entropy falls by 0.34 J K^{-1}. The entropy of a cup (200 ml) of boiling water—it can be calculated by a slightly more involved procedure—is about 200 J K^{-1} higher than at room temperature.

Now we are ready to express the second law in terms of the entropy and to show that a single statement captures the Kelvin and Clausius statements. We begin by proposing the following as a statement of the second law:

> the entropy of the universe increases in the course of any spontaneous change.

The key word here is *universe*: it means, as always in thermodynamics, the system together with its surroundings. There is no prohibition of the system or the surroundings *individually* undergoing a decrease in entropy provided that there is a compensating change elsewhere.

To see that Kelvin's statement is captured by the entropy statement, we consider the entropy changes in the two parts of a heat engine that has no cold sink (Figure 11). When heat leaves the hot source, there is a decrease in the entropy of the system. When that energy is transferred to the surroundings as work, there is no change in the entropy because changes in entropy are defined in terms of the heat transferred, not the work that is done. We shall understand that point more fully later, when we turn to the molecular nature of entropy. There is no other change. Therefore, the overall change is a decrease in the entropy of the universe, which is contrary to the second law. It follows that an engine with no cold sink cannot produce work.

To see that an engine with a cold sink can produce work, we think of an actual heat engine. As before, there is a decrease in entropy when energy leaves the hot sink as heat and there is no change in entropy when some of that heat is converted into work. However,

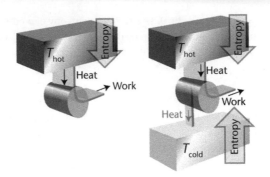

11. An engine like that denied by Kelvin's statement (left) implies a reduction in entropy and is not viable. On the right is shown the consequence of providing a cold sink and discarding some heat into it. The increase in entropy of the sink may outweigh the reduction of entropy of the source, and overall there is an increase in entropy. Such an engine is viable

provided we do not convert all the energy into work, we can discard some into the cold sink as heat. There will now be an increase in the entropy of the cold sink, and provided its temperature is low enough—that is, it is a quiet enough library—even a small deposit of heat into the sink can result in an increase in its entropy that cancels the decrease in entropy of the hot source. Overall, therefore, there can be an increase in entropy of the universe, but only provided there is a cold sink in which to generate a positive contribution. That is why the cold sink is the crucial part of a heat engine: entropy can be increased only if the sink is present, and the engine can produce work from heat only if overall the process is spontaneous. It is worse than useless to have to drive an engine to make it work!

It turns out, as may be quite readily shown, that the fraction of energy withdrawn from the hot source that must be discarded into the cold sink, and which therefore is not available for converting into work, depends only on the temperatures of the source and sink. Moreover, the minimum energy that must be discarded, and therefore the achievement of the maximum efficiency of conversion of heat into work, is given precisely by Carnot's formula. Suppose q leaves the hot source as heat: the entropy falls by q/T_{source}. Suppose q' is discarded into the cold sink: the entropy increases by q'/T_{sink}. For the overall change in entropy to be positive, the minimum amount of heat to discard is such that $q'/T_{sink} = q/T_{source}$, and therefore $q' = qT_{sink}/T_{source}$. That means that the maximum amount of work that can be done is $q - q'$, or $q(1 - T_{sink}/T_{source})$. The efficiency is this work divided by the heat supplied (q), which gives *efficiency* $= 1 - T_{sink}/T_{source}$, which is Carnot's formula.

Now consider the Clausius statement in terms of entropy. If a certain quantity of energy leaves the cold object as heat, the entropy decreases. This is a large decrease, because the object is cold—it is a quiet library. The same quantity of heat enters the hot object. The entropy increases, but because the temperature is higher—the object is a busy street—the resulting increase in entropy is small, and certainly smaller than the decrease in entropy of the cold object. Overall, therefore, there is a decrease in entropy, and the process is not spontaneous, exactly as Clausius's statement implies.

Thus, we see that the concept of entropy captures the two equivalent phenomenological statements of the second law and acts as the signpost of spontaneous change. The first law and the internal energy identify the *feasible* change among all conceivable changes: a process is feasible only if the total energy of the universe remains the same. The second law and entropy identify the *spontaneous* changes among these feasible changes: a feasible process is spontaneous only if the total entropy of the universe increases.

It is of some interest that the concept of entropy greatly troubled the Victorians. They could understand the conservation of energy, for they could presume that at the Creation God had endowed the world with what He would have judged infallibly as exactly the right amount, an amount that would be appropriate for all time. What were they to make of entropy, though, which somehow seemed to increase ineluctably. Where did this entropy spring from? Why was there not an exact, perfectly and eternally judged amount of the God-given stuff?

To resolve these matters and to deepen our understanding of the concept, we need to turn to the molecular interpretation of entropy and its interpretation as a measure, in some sense, of disorder.

Images of disorder

With entropy as a measure of disorder in mind, the change in entropy accompanying a number of processes can be predicted quite simply, although the actual numerical change takes more effort to calculate than we need to display in this introduction. For example, the isothermal (constant temperature) expansion of a gas distributes its molecules and their constant energy over a greater volume, the system is correspondingly less ordered in the sense that we have less chance of predicting successfully where a particular molecule and its energy will be found, and the entropy correspondingly increases.

A more sophisticated way of arriving at the same conclusion, and one that gives a more accurate portrayal of what 'disorder' actually means, is to think of the molecules as distributed over the energy levels characteristic of particles in a box-like region. Quantum mechanics can be used to calculate these allowed energy levels (it boils down to computing the wavelengths of the standing waves that can fit between rigid walls, and then interpreting the wavelengths as energies). The central result is that as the walls of the box are moved apart, the energy levels fall and become less

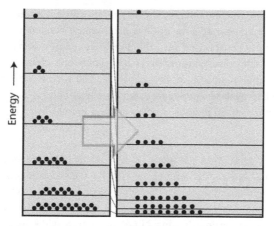

12. **The increase in entropy of a collection of particles in an expanding box-like region arises from the fact that as the box expands, the allowed energies come closer together. Provided the temperature remains the same, the Boltzmann distribution spans more energy levels, so the chance of choosing a molecule from one level in a blind selection decreases. That is, the disorder and the entropy increase as the gas occupies a greater volume**

widely separated (Figure 12). At room temperature, billions of these energy levels are occupied by the molecules, the distribution of populations being given by the Boltzmann distribution characteristic of that temperature. As the box expands, the Boltzmann distribution spreads over more energy levels and it becomes less probable that we can specify which energy level a molecule would come from if we made a blind selection of molecules. This increased uncertainty of the precise energy level a molecule occupies is what we really mean by the 'disorder' of the system, and corresponds to an increased entropy.

A similar picture accounts for the change in entropy as the temperature of a gaseous sample is raised. A simple calculation in classical thermodynamics based on Clausius's definition leads us to expect an increase in entropy with temperature. That increase in molecular terms can be understood, because as the temperature increases at constant volume, the Boltzmann distribution acquires a longer tail, corresponding to the occupation of a wider range of energy levels. Once again, the probability that we can predict which energy level a molecule comes from in a blind selection corresponds to an increase in disorder and therefore to a higher entropy.

This last point raises the question of the value of the entropy at the absolute zero of temperature (at $T = 0$). According to the Boltzmann distribution, at $T = 0$ only the lowest state (the 'ground state') of the system is occupied. That means that we can be absolutely certain that in a blind selection we will select a molecule from that single ground state: there is no uncertainty in the distribution of energy, and the entropy is zero.

These considerations were put on a quantitative basis by Ludwig Boltzmann, who proposed that the so-called *absolute entropy* of any system could be calculated from a very simple formula:

$$S = k \log W$$

The constant k is Boltzmann's constant, which we encountered in Chapter 1 in the relation between β and T, namely $\beta = 1/kT$, and appears here simply to ensure that changes in entropy calculated from this equation have the same numerical value as those calculated from Clausius's expression. Of much greater significance is the quantity W, which is a measure of the number of ways that the molecules of a system can be arranged to achieve the same total energy (the 'weight' of an arrangement). This expression is much harder to implement than the classical thermodynamic expression, and really belongs to the domain of statistical thermodynamics, which is not the subject of this volume. Suffice it to say that Boltzmann's formula can be used to calculate both the absolute entropies of substances, especially if they have simple structures, like a gas, and changes in entropy that accompany various changes, such as expansion and heating. In all cases, the expressions for the changes in entropy correspond exactly to those deduced from Clausius's definition, and we can be confident that the classical entropy and the statistical entropy are the same.

It is an incidental footnote of a personal history that the equation $S = k \log W$ is inscribed on Boltzmann's tombstone as his wonderful epitaph, even though he never wrote down the equation explicitly (it is due to Max Planck). He deserves his constant even if we do not.

Degenerate solids

There are various little wrinkles in the foregoing about which we now need to own up. Because the Clausius expression tells us only the *change* in entropy, it allows us to measure the entropy of a substance at room temperature relative to its value at $T = 0$. In many cases the value calculated for room temperature corresponds within experimental error to the value calculated from Boltzmann's formula using data about molecules obtained from spectroscopy, such as bond lengths and bond angles. In some cases, however, there is a discrepancy, and the thermodynamic entropy differs from the statistical entropy.

We have assumed without comment that there is only one state of lowest energy; one ground state, in which case $W = 1$ at $T = 0$ and the entropy at that temperature is zero. That is, in the technical parlance of quantum mechanics, we assumed that the ground state was 'non-degenerate'. In quantum mechanics, the term 'degeneracy', another hijacked term, refers to the possibility that several different states (for instance, planes of rotation or direction of travel) correspond to the same energy. In some cases, that is not true, and in such cases there may be many different states of the system corresponding to the lowest energy. We could say that the ground states of these systems are highly *degenerate* and denote the number of states that correspond to that lowest energy as D. (I give a visualizable example in a moment.) If there are D such states, then even at absolute zero we have only 1 chance in D of predicting which of these degenerate states a molecule will come from in a blind selection. Consequently, there is disorder in the system even at $T = 0$ and its entropy is not zero. This non-zero entropy of a degenerate system at $T = 0$ is called the *residual entropy* of the system.

Solid carbon monoxide provides one of the simplest examples of residual entropy. A carbon monoxide molecule, CO, has a highly uniform distribution of electric charge (technically, it has only a very tiny electric dipole moment), and there is little difference in energy if in the solid the molecules lie ...CO CO CO..., or ...CO OC CO..., or any other random arrangement of direction. In other words, the ground state of a solid sample of carbon monoxide is highly degenerate. If each molecule can lie in one of two directions, and there are N molecules in the sample, then $D = 2^N$. Even in 1 g of solid carbon monoxide there are 2×10^{22} molecules, so this degeneracy is far from negligible! (Try calculating the value of D.) The value of the residual entropy is $k \log D$, which works out to 0.21 J K^{-1} for a 1 g sample, in good agreement with the value inferred from experiment.

Solid carbon monoxide might seem to be a somewhat rarefied example and of little real interest except as a simple illustration. There is one common substance, though, of considerable importance that is also highly degenerate in its ground state: ice. We do not often think—perhaps ever—of ice being a degenerate solid, but it is, and the degeneracy stems from the location of the hydrogen atoms around each oxygen atom.

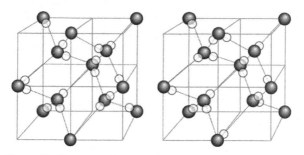

13. The residual entropy of water, reflecting its 'degeneracy' at $T = 0$, arises from the variation in the locations of hydrogen atoms (the small white spheres) between oxygen atoms (the shaded spheres). Although each oxygen atom is closely attached to two hydrogen atoms and makes a more distant link to a hydrogen atom of each of two neighbouring water molecules, there is some freedom in the choice of which links are close and which are distant. Two of the many arrangements are shown here

Figure 13 shows the origin of ice's degeneracy. Each water molecule is H_2O, with two short, strong O–H bonds at about 104° to each other. The molecule is electrically neutral overall, but the electrons are not distributed uniformly, and each oxygen atom has patches of net negative charge on either side of the molecule, and each hydrogen atom is slightly positively charged on account of the withdrawal of electrons from it by the electron-hungry oxygen atom. In ice, each water molecule is surrounded by others in a tetrahedral arrangement, but the slightly positively charged hydrogen atoms of one molecule are attracted to one of the patches of slight negative charge on the oxygen atom of a neighbouring water molecule. This link between molecules is called a *hydrogen bond*, and is denoted O–H$\cdots$O. The link is responsible for the residual entropy of ice, because there is a randomness in whether any particular link is O–H$\cdots$O or O$\cdots$H–O. Each water molecule must have two short O–H bonds (so that it is recognizable as an H_2O molecule), and two H$\cdots$O links to two neighbours, but which two are short and which two are long is almost random. When the statistics of this variability is analysed, it turns out that the residual entropy of 1 g of ice should be 0.19 J K^{-1}, in good agreement with the value inferred from experiment.

Refrigerators and heat pumps

The concept of entropy is the foundation of the operation of heat engines, heat pumps, and refrigerators. We have already seen that a heat engine works because heat is deposited in a cold sink and generates disorder there that compensates, and in general more than compensates, for any reduction in entropy due to the extraction of energy as heat from the hot source. The efficiency of a heat engine is given by the Carnot expression. We see from that expression that the greatest efficiency is achieved by working with the hottest possible source and the coldest possible sink. Therefore, in a steam engine, a term that includes steam turbines as well as classical piston-based engines, the greatest efficiency is achieved by using superheated steam. The fundamental reason for that design feature is that the high temperature of the source minimizes the entropy reduction of the withdrawal of heat (to go unnoticed, it is best to sneeze in a very busy street), so that least entropy has to be generated in the cold sink to compensate for that decrease, and therefore that more energy can be used to do the work for which the engine is intended.

A *refrigerator* is a device for removing heat from an object and transferring that heat to the surroundings. This process does not occur spontaneously because it corresponds to a reduction in total entropy. Thus, when a given quantity of heat is removed from a cool body (a quiet library, in our sneeze analogy), there is a large decrease in entropy. When that heat released into warmer surroundings, there is an increase in entropy, but the increase is smaller than the original decrease because the temperature is higher (it is a busy street). Therefore, overall there

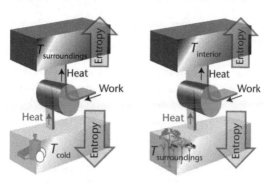

14. The processes involved in a refrigerator and a heat pump. In a refrigerator (left), the entropy of the warm surroundings is increased by at least the amount by which the entropy of the system (the interior of the refrigerator) is decreased; this increase is achieved by adding to the flow of energy by doing work. In a heat pump (right), the same net increase in entropy is achieved, but in this case the interest lies in the energy supplied to the interior of the house

is a net decrease in entropy. We used the same argument in the discussion of Clausius's statement of the second law, which applies directly to this arrangement. A crude restatement of Clausius's statement is that refrigerators don't work unless you turn them on.

In order to achieve a net increase of entropy, we must release more energy into the surroundings than is extracted from the cool object (we must sneeze more loudly in the busy street). To achieve that increase, we must add to the flow of energy. This we can do by doing work on the system, for the work we do adds to the energy stream (Figure 14). When we do work the original energy extracted from the cool body is augmented to heat + work, and that total energy is released into the warmer surroundings. If enough work is done on the system, the release of a large amount of energy into the warm surroundings gives a large increase in entropy and overall there is a net increase in entropy and the process can occur. Of course, to generate the work to drive the refrigerator, a spontaneous process must occur elsewhere, as in a distant power station.

The efficiency of refrigeration is reported as the 'coefficient of performance' of the arrangement. This quantity is defined as the ratio of the heat removed from a cool object to the work that must be done in order to achieve that transfer. The higher the coefficient of performance the less work we have to do to achieve a given transfer—the less power we have to draw from the supply, so the more efficient the refrigerator. By a calculation very similar to that on p. 51, we can conclude that the best coefficient of performance that can be achieved by any arrangement when the object (the food) to be cooled is at a temperature T_{cold} and the surroundings (the kitchen) is at $T_{surroundings}$ is

$$\text{Coefficient of performance (refrigerator)} = \frac{1}{\frac{T_{surroundings}}{T_{cold}} - 1}$$

For instance, if the cold object is cold water at 0°C (273 K) and the refrigerator is in a room at 20°C (293 K), then the coefficient of performance is 14, and to remove 10 kJ of energy from the freezing water, which is enough to freeze about 30 g of the water to ice, under ideal conditions we need to do about 0.71 kJ of work. Actual refrigerators are much less efficient than this thermodynamic limit, not least because heat leaks in from outside and not all the energy supplied to do work joins the energy stream. Air conditioning is essentially refrigeration, and this calculation indicates why it is so expensive—and environmentally damaging—to run. It takes a lot of energy to fight Nature when she wields the second law.

When a refrigerator is working, the energy released into the surroundings is the sum of that extracted from the cooled object and that used to run the apparatus. This remark is the basis of the operation of a *heat pump*, a device for heating a region (such as the interior of a house) by pumping heat from the outside into the interior. A heat pump is essentially a refrigerator, with the cooled object the outside world and the heat transfer arranged to be in the region to be heated. That is, our interest is in the back of the refrigerator, not its interior. The coefficient of performance of a heat pump is defined as the ratio of the total energy released as heat into the region to be heated (at a temperature T_{interior}), to the work done in order to achieve that release. By the same type of calculation as already done for the Carnot efficiency (and which in this case is left to the reader), it turns out that the theoretical best coefficient of performance when the region from which the heat is extracted is at a temperature $T_{\text{surroundings}}$ is

$$\text{Coefficient of performance (heat pump)} = \frac{1}{1 - \dfrac{T_{\text{surroundings}}}{T_{\text{interior}}}}$$

Therefore, if the region to be heated is at 20°C (293 K) and the surroundings are at 0°C (273 K), the coefficient of performance is 15. Thus, to release 1000 J into the interior, we need do only 67 J of work. In other words, a heat pump rated at 1 kW behaves like a 15 kW heater.

Abstracting steam engines

We began this chapter by asserting that we are all steam engines. With a sufficiently abstract interpretation of 'steam engine', that is most definitely true. Wherever structure is to be conjured from disorder, it must be driven by the generation of greater disorder elsewhere, so that there is a net increase in disorder of the universe, with disorder understood in the sophisticated manner that we have sketched. That is clearly true for an actual heat engine, as we have seen. However, it is in fact universally true.

For instance, in an internal combustion engine, the combustion of a hydrocarbon fuel results in the replacement of a compact liquid by a mixture of gases that occupies a volume over 2000 times greater (and still 600 times greater if we allow for the oxygen consumed). Moreover, energy is released by the combustion, and that energy disperses into the surroundings. The design of the engine captures this dispersal in disorder and uses it to build, for instance, a structure from a less ordered pile of bricks, or drive an electric current (an orderly flow of electrons) through a circuit.

The fuel might be food. The dispersal that corresponds to an increase in entropy is the metabolism of the food and the dispersal of energy and matter that that metabolism releases. The structure that taps into that dispersal is not a mechanical chain of pistons and gears, but the biochemical pathways within the body. The structure that those pathways cause to emerge may be proteins assembled from individual amino acids. Thus, as we eat, so we grow. The structures may be of a different kind: they may be works of art. For another structure that can be driven into existence by coupling to the energy released by ingestion and digestion consists of organized electrical activity within the brain constructed from random electrical and neuronal activity. Thus, as we eat, we create: we create works of art, of literature, and of understanding.

The steam engine, in its abstract form as a device that generates organized motion (work) by drawing on the dissipation of energy, accounts for all the processes within our body. Moreover, that great steam engine in the sky, the Sun, is one of the great fountains of construction. We all live off the spontaneous dissipation of its energy, and as we live so we spread disorder into our surroundings: we could not survive without our surroundings. In his seventeenth century meditation, John Donne was unknowingly expressing a version of the second law when he wrote, two centuries before Carnot, Joule, Kelvin, and Clausius, that no man is an island.

Chapter 4
Free energy

The availability of work

Free energy? Surely not! How can energy be free? Of course, the answer lies in a technicality. By *free energy* we do not mean that it is monetarily free. In thermodynamics, the freedom refers to the energy that is free to do work rather than just tumble out of a system as heat.

We have seen that when a combustion occurs at constant pressure, the energy that may be released as heat is given by the change of enthalpy of the system. Although there may be a change in internal energy of a certain value, the system in effect has to pay a tax to the surroundings in the sense that some of that change in internal energy must be used to drive back the atmosphere in order to make room for the products. In such a case, the energy that can be released as heat is less than the change in internal energy. It is also possible for there to be a tax refund in the sense that if the products of a reaction occupy less volume than the reactants, then the system can contract. In this case, the surroundings do work on the system, energy is transferred into it, and the system can release more heat than is given by the change in internal energy: the system recycles the incoming work as outgoing heat. The enthalpy, in short, is an accounting tool for heat that takes into account automatically the tax payable or repayable as work and lets us calculate the heat output without having to calculate the contributions of work in a separate calculation.

The question that now arises is whether a system must pay a tax to the surroundings in order to produce work. Can we extract the full change in internal energy as work, or must some of that change be transferred to the surroundings as heat, leaving less to be used to do work? Must there be a tax, in the form of heat, that a system has to pay in order to do work? Could there even be a tax refund in the sense that we can extract more work than the change in internal energy leads us to expect? In short, by analogy with the role of enthalpy, is there a thermodynamic property that instead of focusing on the net heat that a process can release focuses on the net work instead?

We found the appropriate property for heat, the enthalpy, by considering the first law. We shall find the appropriate property for work by considering the second law and entropy, because a process can do work only if it is spontaneous: non-spontaneous processes have to be driven by doing work, so they are worse than useless for producing work.

To identify spontaneous processes we must note the crucially important aspect of the second law that it refers to the entropy of the *universe*, the sum of the entropies of the system and the surroundings. According to the second law, a spontaneous change is accompanied by an increase in entropy of the *universe*. An important feature of this emphasis on the universe is that a process may be spontaneous, and work producing, even though it is accompanied by a decrease in entropy of the system provided that a greater increase occurs in the surroundings and the *total* entropy increases. Whenever we see the apparently spontaneous reduction of entropy, as when a structure emerges, a crystal forms, a plant grows, or a thought emerges, there is always a greater increase in entropy elsewhere than accompanies the reduction in entropy of the system.

To assess whether a process is spontaneous and therefore capable of producing work, we have to assess the accompanying entropy changes in both the system of interest and the surroundings. It is inconvenient to have to do two separate calculations, one for the system and one for the surroundings. Provided we are prepared to restrict our interest to certain types of change, there is a way to combine the two calculations into one and to carry out the calculation by focusing on the properties of the system alone. By proceeding in that way, we shall be able to identify the thermodynamic property that we can use to assess the work that can be extracted from a process without having to calculate the 'heat tax' separately.

Introducing the Helmholtz energy

The clever step is to realize that if we limit changes to those taking place at constant volume and temperature, then the change in entropy of the surroundings can be expressed in terms of the change in internal energy of the system. That is because at constant volume, the only way that the internal energy can change in a closed system is to exchange energy as heat with the surroundings, and that heat can be used to calculate the change in entropy of the surroundings by using the Clausius expression for the entropy.

When the internal energy of a constant-volume, closed system changes by ΔU, the whole of that change in energy must be due to a heat transaction with the surroundings. If there is an increase in internal energy of the system (for instance, if ΔU = +100 J), then heat equal to ΔU (that is, 100 J) must flow in from the surroundings. The surroundings lose that amount of energy as heat, and so their entropy changes by $-\Delta U/T$, a decrease. If there is a decrease in internal energy of the system, ΔU is negative (for instance, if ΔU = −100 J) and an equal amount of heat (in this case, 100 kJ) flows into the surroundings. Their entropy therefore increases by $-\Delta U/T$ (this is a positive quantity because ΔU is negative when U decreases). In either case, therefore, the total change in entropy of the universe is ΔS (total) $= \Delta S - \Delta U/T$, where ΔS is the change in entropy of the system. This expression is in terms of the properties of the system alone. In a moment we shall use it in the form $-T\Delta S$ (total) $= \Delta U - T\Delta S$, which is obtained by multiplying both sides by $-T$ and changing the order of terms on the right.

To tidy up the calculation, we introduce a combination of the internal energy and the entropy of the system called the *Helmholtz energy*, denoted A, and defined as $A = U - TS$. The German physiologist and physicist Hermann von Helmholtz (1821–1894), after whom this property is named, was responsible for formulating the law of conservation of energy as well as making other major contributions to the science of sensation, colour blindness, nerve propagation, hearing, and thermodynamics in general.

At constant temperature a change in the Helmholtz energy stems from changes in U and S, and $\Delta A = \Delta U - T\Delta S$, exactly as we have just found for $-T\Delta S$(total). So, a change in A is just a disguised form of the change in total entropy of the universe when the temperature and volume of the system are constant. The important implication of this conclusion is that, because spontaneous changes correspond to positive changes (increases) in the total entropy of the universe, provided we limit our attention to processes at constant temperature and volume, spontaneous changes correspond to a decrease in Helmholtz energy *of the system*. The restriction of the conditions to constant temperature and volume has allowed us to express spontaneity solely in terms of the properties of the system: its internal energy, temperature, and entropy.

It probably seems more natural that a spontaneous change corresponds to a decrease in a quantity: in the everyday world, things tend to fall down, not up. However, don't be misled by the seductions of familiarity: the natural tendency of A to decrease is just an artefact of its definition. Because the Helmholtz energy is a disguised version of the total entropy of the universe, the change in direction from 'total entropy up' to 'Helmholtz energy down' simply reflects how A is defined. If you examine the expression for ΔA without its derivation in mind, you will see that a negative value will be obtained if ΔU is negative (a lowering of internal energy of the system) and ΔS is positive. You might then jump to the conclusion that systems tend towards lower internal energy and higher entropy. That would be a wrong interpretation. The fact that a negative ΔU favours spontaneity stems from the fact that it represents the contribution (through $-\Delta U/T$) of the entropy of the surroundings. The *only* criterion of spontaneous change in thermodynamics is the increase in total entropy of the universe.

As well as the Helmholtz energy being a signpost of spontaneous change it has another important role: it tells us the maximum work that can be extracted when a process occurs at constant temperature. That should be quite easy to see: it follows from the Clausius expression for the entropy ($\Delta S = q_{rev}/T$ rearranged into $q_{rev} = T\Delta S$) that $T\Delta S$ is the heat transferred to the surroundings in a reversible process; but ΔU is equal to the sum of the heat and work transactions with the surroundings, and the difference left after allowing for the heat transferred, the value of $\Delta U - T\Delta S$, is the change in energy due to doing work alone. It is for this reason that A is also known as the 'work function' and given the symbol A (because *Arbeit* is the German word for work). More commonly, though, A is called a *free energy*, suggesting that it indicates the energy in a system that is free to do work.

The last point becomes clearer once we think about the molecular nature of the Helmholtz energy. As we saw in Chapter 2, work is uniform motion in the surroundings, as in the moving of all the atoms of a weight in the same direction. The term TS that appears in the definition of $A = U - TS$ has the dimensions of an energy, and can be thought of as a measure of the energy that is stored in a disordered way in the system for which U is the total energy. The difference $U - TS$ is therefore the energy that is stored in an orderly way. We can then think of only the energy stored in an orderly way as being available to cause orderly motion, that is, work, in the surroundings. Thus, only the difference $U - TS$ of the total energy and the 'disordered' energy is energy that is free to do work.

A more precise way of understanding the Helmholtz energy is to think about the significance of changes in its value. Suppose a certain process occurs in a system that causes a change in internal energy ΔU and happens to correspond to a decrease in entropy, so ΔS is negative. The process will be spontaneous and able to produce work only if the entropy of the surroundings increases by a compensating amount, namely ΔS (Figure 15). For that increase to occur, some of the change in internal energy must be released as heat, for only heat transactions result in changes in entropy. To achieve an increase in entropy of magnitude ΔS, according to the Clausius expression, the system must release a quantity of heat of

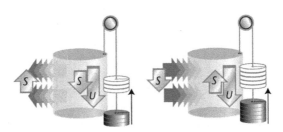

15. On the left a process occurs in a system that causes a change in internal energy ΔU and a decrease in entropy. Energy must be lost as heat to the surroundings in order to generate a compensating entropy there, so less than ΔU can be released as work. On the right, a process occurs with an increase in entropy, and heat can flow in to the system yet still correspond to an increase in total entropy; as a result, more than ΔU can be released as work

magnitude $T\Delta S$. That means that only $\Delta U - T\Delta S$ can be released as work.

According to this discussion, $T\Delta S$ is a tax that the surroundings demand from the system in order to compensate for the reduction in entropy of the system, and only $\Delta U - T\Delta S$ is left for the system to pay out as work. However, suppose the entropy of the system happens to increase in the course of the process. In that case the process is already spontaneous, and no tax need be paid to the surroundings. In fact, it is better than that, because the surroundings can be allowed to supply energy as heat to the system, because they can tolerate a decrease in entropy yet the entropy of the universe will still increase. In other words, the system can receive a tax refund. That influx of energy as heat increases the internal energy of the system and the increase can be used to do more work than in the absence of the influx. That too, is captured by the definition of the Helmholtz energy, for when ΔS is negative, $-T\Delta S$ is a positive quantity and adds to ΔU rather than subtracting from it, and ΔA is bigger than ΔU. In this case, more work can be extracted than we would expect if we considered only ΔU.

Some numbers might give these considerations a sense of reality. When 1 L of gasoline is burned it produces carbon dioxide and water vapour. The change in internal energy is 33 MJ, which tells us that if the combustion takes place at constant volume (in a sturdy, sealed container), then 33 MJ will be released as heat. The change in enthalpy is 0.13 MJ less than the change in internal energy. This figure tells us that if the combustion takes place in a vessel open to the atmosphere, then slightly less (0.13 MJ less, in fact) than 33 MJ will be released as heat. Notice that less heat is released in the second arrangement because 0.13 MJ has been used to drive back the atmosphere to make room for the gaseous products and so less is available as heat. The combustion is accompanied by an increase in entropy because more gas is produced than is consumed (sixteen CO_2 molecules and eighteen

H_2O molecules are produced for every twenty-five O_2 molecules that are consumed, a net increase of nine gas molecules), and it may be calculated that $\Delta S = +8$ kJ K^{-1}. It follows that the change in Helmholtz energy of the system is -35 MJ. Thus, if the combustion took place in an engine, the maximum amount of work that could be obtained is 35 MJ. Note that this is larger than the value of ΔU because the increase in entropy of the system has opened the possibility of heat flowing into the system as a tax refund and there being a corresponding decrease in the surroundings yet leaving the change in total entropy positive. It is, perhaps, refreshing to note that you get a tax refund for every mile you drive; but this is Nature's refund, not the Chancellor's.

Introducing the Gibbs energy

The discussion so far refers to all kinds of work. In many cases we are not interested in expansion work but the work, for example, that can be extracted electrically from an electrochemical cell or the work done by our muscles as we move around. Just as the enthalpy ($H = U + pV$) is used to accommodate expansion work automatically when that is not of direct interest, it is possible to define another kind of free energy that takes expansion work into account automatically and focuses our attention on non-expansion work. The *Gibbs energy*, which is denoted G, is defined as $G = A + pV$. Josiah Willard Gibbs (1839–1903), after whom this property is named, is justifiably regarded as a founding father of chemical thermodynamics. He worked at Yale University throughout his life and was noted for his public reticence. His extensive and subtle work was published in what we now consider to be an obscure journal (*The Transactions of the Connecticut Academy of Science*) and was not appreciated until it was interpreted by his successors.

In the same way as ΔA tells us the total work that a process may do at constant temperature, the change in the Gibbs energy, ΔG, tells us the amount of non-expansion work that a process can do provided the change is taking place at constant temperature and pressure. Just as it is not really possible to give a molecular interpretation of the enthalpy, which is really just a clever accounting device, it is not possible to give a simple explanation of the molecular nature of the Gibbs energy. It is good enough for our purposes to think of it like the Helmholtz energy, as a measure of the energy that is stored in an orderly way and is therefore free to do useful work.

There is another 'just as' to note. Just as a change in the Helmholtz energy is a disguised expression for the change in total entropy of the universe when a process takes place at constant volume and temperature (remember that $\Delta A = -T\Delta S$ (total)), with spontaneous processes characterized by a decrease in A, so the change in Gibbs energy can be identified with a change in total entropy for processes that occur at constant pressure and temperature: $\Delta G = -T\Delta S$ (total). Thus, the criterion of spontaneity of a process at constant pressure and temperature is that ΔG is negative:

at constant volume and temperature, a process is spontaneous if it corresponds to a decrease in Helmholtz energy.

at constant pressure and temperature, a process is spontaneous if it corresponds to a decrease in Gibbs energy.

In each case, the underlying origin of the spontaneity is the increase in entropy of the universe, but in each case we can express that increase in terms of the properties of the system alone and do not have to worry about doing a special calculation for the surroundings.

The Gibbs energy is of the greatest importance in chemistry and in the field of *bioenergetics*, the study of energy utilization in biology. Most processes in chemistry and biology occur at constant temperature and pressure, and so to decide whether they are spontaneous and able to produce non-expansion work we need to consider the Gibbs energy. In fact, when chemists and biologists use the term 'free energy' they almost always mean the Gibbs energy.

The thermodynamics of freezing

There are three applications that I shall discuss here. One is the thermodynamic description of phase transitions (freezing and boiling, for instance; a 'phase' is a form of a given substance, such as the solid, liquid, and vapour phases of water), another is the ability of one reaction to drive another in its non-spontaneous direction (as when we metabolize food in our bodies and then walk or think), and the third is the attainment of chemical equilibrium (as when an electric battery becomes exhausted).

The Gibbs energy of a pure substance decreases as the temperature is raised. We can see how to draw that conclusion from the definition $G = H - TS$, by noting that the entropy of a pure substance is invariably positive. Therefore, as T increases, TS becomes larger and subtracts more and more from H, and G consequently falls. The Gibbs energy of 100 g of liquid water, for instance, behaves as shown in Figure 16 by the line labelled 'liquid'. The Gibbs energy of ice behaves similarly. However, because the entropy of 100 g of ice is lower than that of 100 g of water—because the molecules are more ordered in a solid than the jumble of molecules that constitute a liquid—the Gibbs energy does not fall away as steeply, and is shown by the line labelled 'solid' in the illustration. The entropy of 100 g of water vapour is much greater than that of the liquid because the molecules of a gas occupy a much greater volume and are distributed randomly over it. As a result, the Gibbs energy of the vapour decreases very sharply with increasing temperature, as shown by the line labelled 'gas' in the illustration. At low temperatures we can be confident that the enthalpy of the solid is lower than that of the liquid

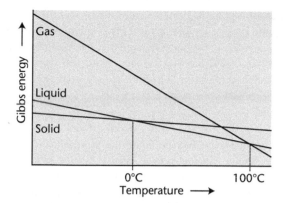

16. The decrease in Gibbs energy with increasing temperature for three phases of a substance. The most stable phase corresponds to the lowest Gibbs energy; thus the solid is most stable at low temperatures, then the liquid, and finally the gas (vapour). If the gas line falls more steeply, it might intersect the solid line before the liquid line does, in which case the liquid is never the stable phase and the solid sublimes directly to a vapour

(because it takes energy to melt a solid) and the enthalpy of the liquid lies below that of the vapour (because it takes energy to vaporize a liquid). That is why we have drawn the Gibbs energies starting in their relative positions on the left of the illustration.

The important feature is that although the Gibbs energy of the liquid is higher than that of the solid at low temperatures, the two lines cross at a particular temperature (0°C, 273 K, as it happens, at normal atmospheric pressure) and from there on the liquid has a lower Gibbs energy than the solid. We have seen that the natural direction of change at constant pressure is to lower Gibbs energy (corresponding, remember, to greater total entropy), so we can infer that at low temperature the solid form of water is the most stable, but that once the temperature reaches 0°C the liquid becomes more stable and the substance spontaneously melts.

The Gibbs energy of the liquid remains the lowest of the three phases until the steeply falling line for the vapour intersects it. For water, at normal atmospheric pressure that intersection occurs at 100°C (373 K), and from that temperature on, the vapour is the most stable form of water. The system spontaneously falls to lower Gibbs energy, so vaporization is spontaneous above 100°C: the liquid boils.

There is no guarantee that the 'liquid' line intersects the 'solid' line before the 'vapour' line has plunged down and crossed the 'solid' line first. In such a case, the substance will make a direct transition from solid to vapour without melting to an inter-mediate liquid phase. This is the process called *sublimation*. Dry ice (solid carbon dioxide) behaves in this way, and converts directly to carbon dioxide gas.

All phase changes can be expressed thermodynamically in a similar way, including melting, freezing, condensation, vaporization, and sublimation. More elaborate discussions also enable us to discuss the effect of pressure on the temperatures at which phase transitions occur, for pressure affects the locations of the lines showing the dependence of Gibbs energy on temperature in different ways, and the intersection points move accordingly. The effect of pressure on the graph lines for water accounts for a familiar example, for at sufficiently low pressure its 'liquid' line does not intersect its 'solid' line before its 'vapour' line has plunged down, and it too sublimes. This behaviour accounts for the disappearance of hoar frost on a winter's morning, when actual ice is truly dry.

Living off Gibbs energy

Our bodies live off Gibbs energy. Many of the processes that constitute life are non-spontaneous reactions, which is why we decompose and putrefy when we die and these life-sustaining reactions no longer continue. A simple (in principle) example is

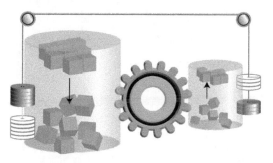

17. A process that corresponds to a large increase in total energy (represented here by an increase in disorder on the left) can drive a process in which order emerges from disorder (on the right). This is analogous to a falling heavy weight being able to raise a lighter weight

the construction of a protein molecule by stringing together in an exactly controlled sequence numerous individual amino acid molecules. The construction of a protein is not a spontaneous process, as order must be created out of disorder. However, if the reaction that builds a protein is linked to a strongly spontaneous reaction, then the latter might be able to drive the former, just as the combustion of a fuel in an engine can be used to drive an electric generator to produce an orderly flow of electrons, an electric current. A helpful analogy is that of a weight which can be raised by coupling it to a heavier weight that raises the lighter weight as it falls (Figure 17).

In biology a very important 'heavy weight' reaction involves the molecule adenosine triphosphate (ATP). This molecule consists of a knobbly group and tail of three alternating phosphorus and oxygen groups of atoms (hence the 'tri' and the 'phosphate' in its name). When a terminal phosphate group is snipped off by reaction with water (Figure 18), to form adenosine diphosphate (ADP), there is a substantial decrease in Gibbs energy, arising in part from the increase in entropy when the group is liberated from the chain. Enzymes in the body make use of this change in Gibbs

75

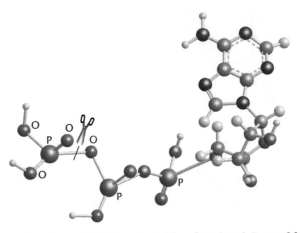

18. A molecular model of adenosine triphosphate (ATP). Some of the phosphorus (P) and oxygen (O) atoms are marked. Energy is released when the terminal phosphate group is severed at the location shown by the line. The resulting ADP molecule must be 'recharged' with a new phosphate group: that recharging is achieved by the reactions involved in digestion and metabolism of food

energy—this falling heavy weight—to bring about the linking of amino acids, and gradually build a protein molecule. It takes the effort of about three ATP molecules to link two amino acids together, so the construction of a typical protein of about 150 amino acid groups needs the energy released by about 450 ATP molecules.

The ADP molecules, the husks of dead ATP molecules, are too valuable just to discard. They are converted back into ATP molecules by coupling to reactions that release even more Gibbs energy—act as even heavier weights—and which reattach a phosphate group to each one. These heavy-weight reactions are the reactions of metabolism of the food that we need to ingest regularly. That food may be the material that has been driven into existence by even heavier reactions—reactions that release even more Gibbs energy, and ultimately off the nuclear processes that occur on the Sun.

Chemical equilibrium

Our final illustration of the utility of the Gibbs energy is one of crucial importance in chemistry. It is a well-known feature of chemical reactions that they all proceed to a condition known as 'equilibrium' in which some reactants (the starting materials) are present and the reaction has appeared to have come to a stop before all the reactants have been converted into products. In some cases the composition corresponding to equilibrium is virtually pure products and the reaction is said to be 'complete'. Nevertheless, even in this case there are one or two molecules of reactants among the myriad product molecules. The explosive reaction of hydrogen and oxygen to form water is an example. On the other hand, some reactions do not appear to go at all. Nevertheless, at equilibrium there are one or two product molecules among the myriad reactant molecules. The dissolution of gold in water is an example. A lot of reactions lie between these extremes, with reactants and products both in abundance, and it is a matter of great interest in chemistry to account for the composition corresponding to equilibrium and how it responds to the conditions, such as the temperature and the pressure. An important point about chemical equilibrium is that when it is achieved the reaction does not simply grind to a halt. At a molecular level all is turmoil: reactants form products and products decompose into reactants, but both processes occur at matching rates, so there is no net change. Chemical equilibrium is *dynamic* equilibrium, so it remains sensitive to the conditions: the reaction is not just lying there dead.

The Gibbs energy is the key. Once again we note that at constant temperature and pressure a system tends to change in the direction corresponding to decreasing Gibbs energy. When applying it to chemical reactions, we need to know that the Gibbs energy of the reaction mixture depends on the composition of the mixture. That dependence has two origins. One is the difference in Gibbs energies of the pure reactants and the pure products: as the composition changes from pure reactants to pure products, so the Gibbs energy changes from one to the other. The second contribution is from the mixing of the reactants and products, which is a contribution to the entropy of the system and therefore, through $G = H - TS$, to the Gibbs energy too. This contribution is zero for pure reactants and for pure products (where there is nothing to mix), and is a maximum when the reactants and products are both abundant and the mixing is extensive.

When both contributions are taken into account, it is found that the Gibbs energy passes through a minimum at an intermediate composition. This composition corresponds to equilibrium. Any composition to the left or right of the minimum has a higher Gibbs energy, and the system tends spontaneously to migrate to lower Gibbs energy and attain the composition corresponding to equilibrium. If the composition is at equilibrium, then the

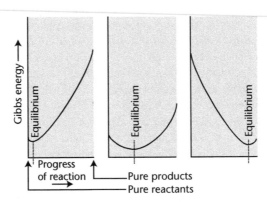

19. The variation of the Gibbs energy of a reaction mixture as it changes from pure reactants to pure products. In each case, the equilibrium composition, which shows no further net tendency to change, occurs at the minimum of the curve

reaction has no tendency to run in either direction. In some cases (Figure 19), the minimum lies far to the left, very close to pure reactants, and the Gibbs function reaches its minimum value after only a few molecules of products are formed (as for gold dissolving in water). In other cases, the minimum lies far to the right, and almost all the reactants must be consumed before the minimum is reached (as for the reaction between hydrogen and oxygen).

One everyday experience of a chemical reaction reaching equilibrium is an exhausted electric battery. In a battery, a chemical reaction drives electrons through an external circuit by depositing electrons in one electrode and extracting them from another electrode. This process is spontaneous in the thermo-dynamic sense, and we can imagine it taking place as the reactants sealed into the battery convert to products, and the composition migrates from left to right in Figure 19. The Gibbs energy of the system falls, and in due course reaches its minimum value. The chemical reaction has reached equilibrium. It has no further tendency to change into products, and therefore no further tendency to drive electrons through the external circuit. The reaction has reached the minimum of its Gibbs energy and the battery—but not the reactions still continuing inside—is dead.

Chapter 5
The third law

The unattainability of zero

I have introduced the temperature, the internal energy, and the entropy. Essentially the whole of thermodynamics can be expressed in terms of these three quantities. I have also introduced the enthalpy, the Helmholtz energy, and the Gibbs energy; but they are just convenient accounting quantities, not new fundamental concepts. The third law of thermodynamics is not really in the same league as the first three, and some have argued that it is not a law of thermodynamics at all. For one thing, it does not inspire the introduction of a new thermodynamic function. However, it does make possible their application.

Hints of the third law are already present in the consequences of the second law, where we considered its implications for refrigeration. We saw that the coefficient of performance of a refrigerator depends on the temperature of the body we are seeking to cool and that of the surroundings. The coefficient of performance falls to zero as the temperature of the cooled body approaches zero. That is, we need to do an ever increasing, and ultimately infinite, amount of work to remove energy from the body as heat as its temperature approaches absolute zero.

There is another hint about the nature of the third law in our discussion of the second. We have seen that there are two approaches to the definition of entropy, the thermodynamic, as expressed in Clausius's definition, and the statistical, as expressed by Boltzmann's formula. They are not quite the same: the thermodynamic definition is for *changes* in entropy; the statistical definition is an *absolute* entropy. The latter tells us that a fully ordered system, one without positional disorder and without thermal disorder—in short, a system in its nondegenerate ground state—has zero entropy regardless of the chemical composition of the substance, but the former leaves open the possibility that the entropy has a value other than zero at $T = 0$ and that different substances have different entropies at that temperature.

The third law is the final link in the confirmation that Boltzmann's and Clausius's definitions refer to the same property and therefore justifies the interpretation of entropy changes calculated by using thermodynamics as changes in disorder of the system, with disorder understood to have the slightly sophisticated interpretation discussed in Chapter 3. It also makes it possible to use data obtained by thermal measurements, such as heat capacities, to predict the composition of reacting systems that correspond to equilibrium. The third law also has some troublesome implications, especially for those seeking very low temperatures.

Extreme cold

As usual in classical thermodynamics, we focus on observations made outside the system of interest, in its surroundings, and close our minds, initially at least, to any knowledge or preconceptions we might have about the molecular structure of the system. That is, to establish a law of classical thermodynamics, we proceed wholly phenomenologically.

Interesting things happen to matter when it is cooled to very low temperatures. For instance, the original version of *superconductivity*, the ability of certain substances to conduct electricity with zero resistance, was discovered when it became possible to cool matter to the temperature of liquid helium (about 4 K). Liquid helium itself displays the extraordinary property of *superfluidity*, the ability to flow without viscosity and to creep over the apparatus that contains it, when it is cooled to about 1 K. The challenge, partly because it is there, is to cool matter to absolute zero itself. Another challenge, to which we shall return, is to explore whether it is possible—and even meaningful—to cool matter to temperatures below absolute zero of temperature; to break, as it were, the temperature barrier.

Experiments to cool matter to absolute zero proved to be very difficult, not merely because of the increasing amount of work that has to be done to extract a given amount of heat from an object as its temperature approaches zero. In due course, it was conceded that it is impossible to attain absolute zero using a conventional thermal technique; that is, a refrigerator based on the heat engine design we discussed in Chapter 3. This empirical observation is the content of the phenomenological version of the third law of thermodynamics:

> no finite sequence of cyclic processes can succeed in cooling a body to absolute zero.

This is a negative statement; but we have seen that the first and second laws can also be expressed as denials (no change in internal energy occurs in an isolated system, no heat engine operates without a cold sink, and so on), so that is not a weakening of its implications. Note that it refers to a *cyclic* process: there might be other kinds of process that can cool an object to absolute zero, but the apparatus that is used will not be found to be in the same state as it was initially.

You will recall that in Chapter 1 we introduced the quantity β as a more natural measure of temperature (with $\beta = 1/kT$), with absolute zero corresponding to infinite β. The third law as we have stated it, transported to a world where people use β to express temperature, appears almost self-evident, for it becomes 'no finite sequence of cyclic processes can succeed in cooling a body to infinite β', which is like saying that no finite ladder can be used to reach infinity. There must be more to the third law than appearances suggest.

Achieving zero

We have remarked that thermodynamicists become excited when nothing at all happens and that negations can have seriously positive consequences, provided we think about the consequences carefully. The pathway to a positive implication in this case is entropy, and we need to consider how the third law impinges on the thermodynamic definition of entropy. To do so, we need to think about how low temperatures are achieved.

Let's suppose that the system consists of molecules that each possess one electron. We need to know that a single electron has the property of *spin*, which for our purposes we can think of as an actual spinning motion. For reasons rooted in quantum mechanics, an electron spins at a fixed rate and may do so either clockwise or anticlockwise with respect to a given direction. These two spin states are denoted ↑ and ↓. The spinning motion of the electron gives rise to a magnetic field, and we may think of each electron as behaving like a tiny bar magnet oriented in either of two directions. In the presence of an applied magnetic field, the two orientations of the bar magnets arising from the two spin states have different energies, and the Boltzmann distribution can be used to calculate the small difference in populations for a given temperature. At room temperature there will be slightly more lower energy ↓ spins than higher energy ↑ spins. If somehow we could contrive to convert some of the ↑ into ↓ spins, then the population difference will correspond to a lower temperature, and we shall have cooled the sample. If we could contrive to make all the spins ↓, then we shall have reached absolute zero.

We shall represent the sample at room temperature and in the absence of a magnetic field by ... ↓↓↑↓↑↑↓↓↓↑↓ ... with a random distribution of ↓ and ↑ spins. These spins are in thermal contact with the rest of the material in the sample and share the same temperature. Now we increase the magnetic field with the sample in thermal contact with its surroundings. Because the sample can give up energy to its surroundings, the electron spin populations can adjust. The sample becomes ... ↑↓↓↑↓↓↓↑↑↓↑ ... with a small preponderance of ↓ spins over ↑ spins. The spin arrangement contributes to the entropy, and so we can conclude that, because the spin distribution is less random than it was initially (because we can be more confident about getting a ↓ in a blind selection), the entropy of the sample has been reduced (Figure 20). That is, by turning up the magnetic field and allowing energy to escape as the electron spins realign, we lower the entropy of the sample.

Now consider what happens when we isolate the sample thermally from its surroundings and gradually reduce the applied field to zero. A process that occurs without the transfer of energy as heat is called adiabatic, as we saw in Chapter 1, so this step is the 'adiabatic demagnetization' step that gives the process its name. Because the process is adiabatic the entropy of the entire sample (the spins and their immediate surroundings) remains the same. The electron spins no longer have a magnetic field to align against, so they resume their original higher entropy random arrangement like ... ↓↓↑↓↑↑↓↓↓↑↓ .... However, because there is no change in the overall entropy of the sample, the entropy of the molecules that carry the electrons must be lowered, which corresponds to a lowering of temperature. Isothermal magnetization followed by adiabatic demagnetization has cooled the sample.

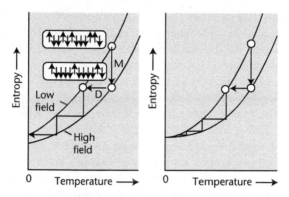

20. The process of adiabatic demagnetization for reaching low temperatures. The arrows depict the spin alignment of the electrons in the sample. The first step (M) is isothermal magnetization, which increases the alignment of the spins, the second step (D) is adiabatic demagnetization, which preserves the entropy and therefore corresponds to a lowering of temperature. If the two curves did not meet at $T = 0$, it would be possible to lower the temperature to zero (as shown on the left). That a finite sequence of cycles does not bring the temperature to zero (as shown on the right) implies that the curves meet at $T = 0$

Next, we repeat the process. We magnetize the newly cooled sample isothermally, isolate it thermally, and reduce the field adiabatically. This cycle lowers the temperature of the sample a little more. In principle, we can repeat this cyclic process, and gradually cool the sample to any desired temperature.

At this point, however, the wolf inside the third law hurls off its sheep's clothing. If the entropy of the substance with and without the magnetic field turned on were to be like that shown in the left-hand half of Figure 20, then we could select a series of cyclic changes that would bring the sample to $T = 0$ in a finite series of steps. It has not proved possible to achieve absolute zero in this way. The implication is that the entropy does not behave as shown on the left, but must be like that shown on the right of the illustration, with the two curves coinciding at $T = 0$.

There are other processes that we might conceive of using to reach absolute zero in a cyclic manner. For instance, we could take a gas and compress it isothermally then allow it to expand adiabatically to its initial volume. The adiabatic expansion of a gas does work, and as no heat enters the system, the internal energy falls. As we have seen, the internal energy of a gas arises largely from the kinetic energy of its molecules, so adiabatic expansion must result in their slowing down and therefore to a lowering of the temperature. At first sight, we might expect to repeat this cycle of isothermal compression and adiabatic expansion, and hope to bring the temperature down to zero. However, it turns out that the effect of adiabatic expansion on the temperature diminishes as the temperature falls, so the possibility of using this technique is thwarted.

An even more elaborate technique would involve a chemical reaction in which the process involved using a reactant A to form a product B, finding an adiabatic path to recreate A, and continuing this cycle. Once again, careful analysis shows that the technique will fail to reach absolute zero because the entropies of A and B converge on the same value as the temperature approaches zero.

The common feature of this collective failure is traced to the convergence of the entropies of substances to a common value as T approaches zero. So, we can replace the phenomenological statement of the third law with a slightly more sophisticated version expressed in terms of the entropy:

> the entropy of every pure, perfectly crystalline substance approaches the same value as the temperature approaches zero.

Note that the experimental evidence and the third law do not tell us the absolute value of the entropy of a substance at $T = 0$. All the law implies is that all substances have the same entropy at $T = 0$ provided they have nondegenerate ground states—no residual order arising from positional disorder of the type characteristic of ice. However, it is expedient and sensible to choose the common value for the entropy of all perfectly crystalline substances as zero, and thus we arrive at the conventional 'entropy' statement of the third law:

the entropy of all perfectly crystalline substances is zero at $T = 0$.

The third law does not introduce a new thermodynamic function, and is therefore not the same type of law as the other three: it simply implies that the entropy can be expressed on an absolute scale.

Some technical consequences

At first sight, the third law is important only to that very tiny section of humanity struggling to beat the low-temperature record (which, incidentally, currently stands at 0.000 000 000 1 K for solids and at about 0.000 000 000 5 K for gases—when molecules travel so slowly that it takes 30 s for them to travel an inch). The law would seem to be irrelevant to the everyday world, unlike the other three laws of thermodynamics, which govern our daily lives with such fearsome relevance.

There are indeed no pressing consequences of the third law for the everyday world, but there are serious consequences for those who inhabit laboratories. First, it eliminates one of science's most cherished idealizations, that of a perfect gas. A perfect gas—a fluid that can be regarded as a chaotic swarm of independent molecules in vigorous random motion—is taken to be the starting point for many discussions and theoretical formulations in thermodynamics, but the third law rules out its existence at $T = 0$. The arguments are too technical to reproduce here, but all stem from the vanishing of entropy at $T = 0$. There are technical salves to what might seem fatal injuries to the fabric of thermodynamics, so the subject does survive this onslaught from its own laws. Another technical consequence is that one major application of thermodynamics to chemistry lies in the use of thermal data, specifically heat capacities measured over a range of temperatures, to calculate the equilibrium composition of reactions and thus to decide whether a reaction is likely to be successful or not and to optimize the conditions for its implementation in industry. The third law provides the key to this application, which could not be done if the entropies of substances were different at absolute zero.

Temperatures below zero

Absolute zero is unattainable—in a sense. Too much should not be read into the third law, because in the form that expresses the unattainability of absolute zero it concerns processes that maintain thermal equilibrium and are cyclic. It leaves open the possibility that there are *non-cyclic* processes that can reach absolute zero. The intriguing consequential question that might occur is whether it is possible to contrive special techniques that take a sample to the other side of zero, where the 'absolute' temperature is negative, whatever that means.

To understand what it means for a body to have a temperature below zero, below, paradoxically, its lowest possible value, we need to recall from Chapter 1 that T is a parameter that occurs in the Boltzmann distribution and which specifies the populations of the available energy levels. It will be simplest, and in practice most readily realizable, to consider a system that has only two energy levels, a ground state and a second state above it in energy. An actual example is an electron spin in a magnetic field, of the kind already mentioned in this chapter. As we have already remarked, because these two spin states correspond to opposite orientations of the bar magnet, they have two different energies.

According to the Boltzmann distribution, at all finite temperatures there will be more electrons in the state of lower energy (the ↓ state) than of higher energy (the ↑ state). At $T = 0$, all the electrons will be in the ground state (all will be ↓) and the entropy will be zero. As the temperature is raised, electrons migrate into the upper state, and the internal energy and the entropy both increase. When the temperature becomes infinite, the electrons are distributed equally over the two states, with half the electrons ↓ and the other half ↑. The entropy has reached its maximum value, a value which according to Boltzmann's formula is proportional to log 2.

Note in passing that an infinite temperature does not mean that all the electrons are in the upper state: at infinite temperature, there are equal populations in the two states. This is a general conclusion: if a system has many energy levels, then when the temperature is infinite, all the states are equally populated.

Now suppose that T is negative, such as −300 K. When T is given a negative value in the Boltzmann distribution we find that the population of the upper state is predicted to be greater than that in the lower state. For instance, if it happens that at 300 K the ratio of populations upper : lower is 1 : 5, then setting $T = -300$ K gives a ratio of 5 : 1, with five times as many electron spins in the upper energy state than in the lower state. Setting $T = -200$ K gives a ratio of 11 : 1, and with $T = -100$ K the ratio is 125 : 1. At −10 K the population of the upper state is nearly 1 000 000 000 000 000 000 000 times greater. Notice how, as the temperature approaches zero from below (−300 K, −200 K, −100 K, . . .), the population migrates almost exclusively into the upper state. In fact, just below 0, the population is entirely in the *upper* state. Immediately above zero the population is entirely in the *lower* state. We have seen that as the temperature is raised from zero to infinity, the population migrates from the lower state and the two states become equally populated. As the temperature is lowered from zero to *minus* infinity the population migrates

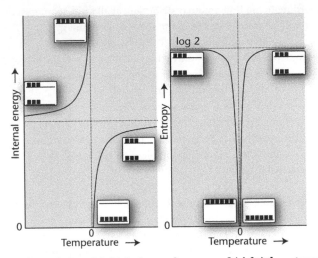

21. The variation of (left) the internal energy and (right) the entropy for a two-level system. The expressions for these two properties can be calculated for negative temperatures, as shown on the left of each illustration. Just above $T = 0$ all the molecules are in the ground state; just below $T = 0$ they are all in the upper state. As the temperature becomes infinite in either direction, the populations become equal

from the upper state into the ground state, and at minus infinity itself the populations are again equal.

We saw in Chapter 1 that the inverse temperature, specifically $\beta = 1/kT$, is a more natural measure of temperature than T itself. That it is to humanity's regret that β has not been adopted becomes very clear when instead of plotting the energy against T as shown in Figure 21, we plot it against β, for in Figure 22 we get a nice smooth curve instead of the unpleasant jump in the first graph at $T = 0$. You should also note that there is a long expanse of β at high β, corresponding to very low temperatures, and it should not be surprising that there is plenty of room for a lot of interesting physics as T approaches zero. We are stuck, however, with the inconvenience of T in place of the smooth convenience of β.

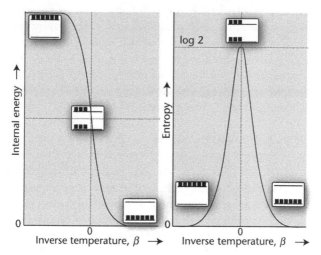

22. The same system as in Figure 21 but plotted against β instead of T. The internal energy varies smoothly across the range

If we could contrive a system in which there are more ↑ (high energy) electrons than ↓ (low energy) electrons, then from the Boltzmann distribution we would ascribe it a negative temperature. Thus, if we could contrive a system in which there are five times as many ↑ electrons as ↓ electrons, then for the same energy separation as we assumed in the preceding discussion, we would report the temperature as −300 K. If we managed to contrive a ratio of 11 : 1, then the temperature would be reported as −200 K, and so on. Note that it is easier to contrive extremely low temperatures (those approaching minus infinity) because they correspond to very tiny imbalances of populations whereas large imbalances correspond to temperatures just below zero. If the temperature is −1 000 000 K, the population ratio is only 1.0005 : 1, a difference of only 0.05 per cent.

The entropy tracks these changes in the distribution of populations. Thus, whereas S increases from zero to log 2 (in suitable units) as T rises from zero to infinity, so too does it increase from zero to log 2 at infinitely negative temperature. On either side of zero we know precisely which state every electron is in (as ↓ just above zero and as ↑ at just below zero), so $S = 0$. At either extreme of infinity, the two states are equally populated, so a random selection gives equal chances of drawing ↑ and ↓. You should reflect on these figures in terms of β instead of T.

The big question is whether the inversion of a thermal equilibrium (that is, Boltzmann) population can be contrived. It can, but not by thermodynamic procedures. There are a variety of experimental techniques available for *polarizing*, as it is called, a collection of electron or nuclear spins that use pulses of radiofrequency energy. In fact, there is an everyday device that makes use of negative temperatures: the laser. The essential principle of a laser is to produce large numbers of atoms or molecules in an excited state and then to stimulate them to discard their energy collectively. What we have referred to as the ↓ and ↑ states of an electron can be regarded as the analogues of the lower and upper energy states of the atom or molecule in the laser material, and the inverted population on which the laser effect depends corresponds to a negative absolute temperature. All the laser-equipped devices we use around the home, as in CD and DVD players, operate at temperatures below zero.

Thermodynamics below zero

The concept of negative temperature really applies in practice only to systems that possess two energy levels. To achieve a distribution of populations over three or more energy levels that can be expressed as a Boltzmann distribution with a formally negative value of the temperature is much more difficult and highly artificial. Moreover, negative temperatures effectively take us outside the domain of classical thermodynamics because they have to be contrived and in general do not persist for more than very short periods. Nevertheless, it is possible—and interesting—to reflect on the thermodynamic properties of systems that have formally negative temperatures.

The first law is robust, and independent of how populations are distributed over the available states. Therefore, in a region of negative temperature, energy is conserved and the internal energy may be changed by doing work or making use of a temperature difference.

The second law survives because the definition of entropy survives, but its implications are different. Thus, suppose energy *leaves* a system as heat at a negative temperature, then according to Clausius's expression the entropy of the system *increases*: the energy change is negative (say −100 J) and so is the temperature (say −200 K), so their ratio is positive (in this case (−100 J)/(−200 K) = +0.5 J K^{-1}). We can understand that conclusion at a molecular level by thinking about a two-level system: think of the inverted population, which has a high energy but low entropy, losing some of its energy and the population returning towards equality, a high entropy (log 2) condition, so the entropy increases as energy is lost. Similarly, if energy as heat enters a system of negative temperature, the entropy of the system decreases (if 100 J enters a system as −200 K, the change in entropy is (+100 J)/(−200 K) = −0.5 J K^{-1}, a decrease). In this case, the upper state becomes more highly populated as energy floods in, so the population moves towards a greater imbalance, towards the entire population being in the upper state and the entropy close to zero.

The second law accounts for the 'cooling' of a system with a negative temperature. Suppose heat leaves the system: its entropy increases (as we have just seen). If that energy enters the surroundings at a positive temperature, their entropy also increases. Therefore, there is an overall increase in entropy when heat is transferred from a region of negative temperature to one of 'normal' positive temperature. Once the populations of the first system have equalized, we can treat the system as having a very high positive temperature—one close to infinite temperature. From this point on, we have an ordinary very hot system in contact with a cooler system, and the entropy continues to increase as heat flows from the former to the latter. In short, the second law implies that there will be a spontaneous transfer of heat from a system of negative temperature in contact with one of positive temperature and that the process will continue until the temperatures of the two systems are equal. The only difference between this discussion and the conventional one is that, provided one system has a negative temperature, the heat flows from the system with the lower (negative) temperature to the one with the higher (positive) temperature.

If both systems have a negative temperature, heat flows from the system with the higher (less negative) temperature to the system with the lower (more negative) temperature. To understand that conclusion, suppose a system at -100 K loses 100 J as heat: the entropy increases by $(-100 \text{ J})/(-100 \text{ K}) = 1 \text{ J K}^{-1}$. If that same heat is deposited in a system at -200 K, the entropy changes by $(+100 \text{ J})/(-200 \text{ K}) = -0.5 \text{ J K}^{-1}$, a decrease. Therefore, overall the total entropy of the two systems increases by 0.5 J K^{-1} and the flow of heat from -100 K (the higher temperature) to -200 K is spontaneous.

The efficiency of a heat engine, which is a direct consequence of the second law, is still defined by the Carnot expression (p. 40). For your convenience again: $\varepsilon = 1 - T_{sink}/T_{source}$. However, if the temperature of the cold reservoir is negative, the efficiency of the engine may be greater than 1. For instance, if the temperature of the hot source is 300 K and that of the cold sink is -200 K, then the efficiency works out as 1.67: we can expect to get more work from the engine than the heat we extract from the hot source. The extra energy actually comes from the cold sink, because, as we have seen, extracting heat from a source with a negative temperature increases its entropy. In a sense, as the inverted population in the cold (negative) sink tumbles back down towards equality, the energy released contributes to the work that the engine produces.

If both the source and the sink of a heat engine are at negative temperatures, the efficiency is less than 1, and the work done is the conversion of the energy withdrawn as heat from the 'warmer', less negative, sink.

The third law requires a slight amendment on account of the discontinuity of the thermal properties of a system across $T = 0$. First, on the 'normal' side of zero, we simply have to change the law to read 'it is impossible in a finite number of cycles to cool any system down to zero.' On the other side of zero, the law takes the form that 'it is impossible in a finite number of cycles to *heat* any system up to zero.' Not, I suspect, that anyone would wish to try!

Conclusion

We are at the end of our journey. We have seen that thermo-dynamics, the study of the transformations of energy, is a subject of great breadth and underlies and elucidates many of the most common concepts of the everyday world, such as temperature, heat, and energy. We have seen that it emerged from reflections on measurements of the properties of bulk samples, but that the molecular interpretation of its concepts enriches our understanding of them.

The first three laws each introduce a property on which the edifice of thermodynamics is based. The zeroth law introduced the concept of temperature, the first law introduced internal energy, and the second law introduced entropy. The first law circumscribed the feasible changes in the universe: those that conserve energy. The second law identified from among those feasible changes the ones that are spontaneous—which have a tendency to occur without us having to do work to drive them. The third law brought the molecular and empirical formulations of thermodynamics into coincidence, uniting the two rivers.

Where I have feared to tread is in two domains that spring from or draw analogies with thermodynamics. I have not touched on the still insecure world of non-equilibrium thermodynamics, where attempts are made to derive laws relating to the rate at which a process produces entropy as it takes place. Nor have I touched on the extraordinary, and understandable, analogies in the field of information theory, where the content of a message is closely related to the statistical thermodynamic definition of entropy. I have not mentioned other features that some regard as central to a deep understanding of thermodynamics, such as the fact that its laws, especially the second law, are statistical in nature and therefore admit to brief failures as molecules fluctuate into surprising arrangements.

What I have sought to cover are the core concepts, concepts that effectively sprang from the steam engine but reach out to embrace the unfolding of a thought. This little mighty handful of laws truly drive the universe, touching and illuminating everything we know.

Further reading

If you would like to take any of these matters further, then here are some suggestions. I wrote about the conservation of energy and the concept of entropy in my *Galileo's Finger: The Ten Great Ideas of Science* (Oxford University Press, 2003), at about this level but slightly less quantitatively. In *The Second Law* (W. H. Freeman & Co., 1997) I attempted to demonstrate that law's concepts and implications largely pictorially, inventing a tiny universe where we could see every atom. More serious accounts will be found in my various textbooks. In order of complexity, these are *Chemical Principles: The Quest for Insight* (with Loretta Jones, W. H. Freeman & Co., 2010), *Elements of Physical Chemistry* (with Julio de Paula, Oxford University Press and W. H. Freeman & Co., 2009), and *Physical Chemistry* (with Julio de Paula, Oxford University Press and W. H. Freeman & Co., 2010).

Others, of course, have written wonderfully about the laws. I can direct you to that most authoritative account, *Thermodynamics*, by G. N. Lewis and M. Randall (McGraw-Hill, 1923; revised by K. S. Pitzer and L. Brewer, 1961). Other useful and reasonably accessible texts on my shelves are *The Theory of Thermodynamics*, by J. R. Waldram (Cambridge University Press, 1985), *Applications of Thermodynamics*, by B. D. Wood (Addison-Wesley, 1982), *Entropy Analysis*, by N. C. Craig (VCH, 1992), *Entropy in Relation to Incomplete Knowledge*, by K. G. Denbigh and J. S. Denbigh (Cambridge University Press, 1985), and *Statistical Mechanics: A Concise Introduction for Chemists*, by B. Widom (Cambridge University Press, 2002).

Index

Symbol and unit index

MOLECULES
A Very Short Introduction
Philip Ball

The processes in a single living cell are akin to that of a city teeming with molecular inhabitants that move, communicate, cooperate, and compete. In this Very Short Introduction, Philip Ball uses a non-traditional approach to chemistry, focusing on what chemistry might become during this century, rather than a survey of its past

He explores the role of the molecule in and around us - how, for example, a single fertilized egg can grow into a multi-celled Mozart, what makes spider's silk insoluble in the morning dew, and how this molecular dynamism is being captured in the laboratory, promising to reinvent chemistry as the central creative science of the century.

'Almost no aspect of the exciting advances in molecular research studies at the beginning of the 21st Century has been left untouched and in so doing, Ball has presented an imaginative, personal overview, which is as instructive as it is enjoyable to read.'
Harry Kroto, Chemistry Nobel Laureate 1996

'A lucid account of the way that chemists see the molecular world . . . the text is enriched with many historical and literature references, and is accessible to the reader untrained in chemistry'
THES, 04/01/2002

http://www.oup.co.uk/isbn/0–19–285430–5

LOGIC
A Very Short Introduction
Graham Priest

Logic is often perceived as an esoteric subject, having little to do with the rest of philosophy, and even less to do with real life. In this lively and accessible introduction, Graham Priest shows how wrong this conception is. He explores the philosophical roots of the subject, explaining how modern formal logic deals with issues ranging from the existence of God and the reality of time to paradoxes of self-reference, change, and probability. Along the way, the book explains the basic ideas of formal logic in simple, non-technical terms, as well as the philosophical pressures to which these have responded. This is a book for anyone who has ever been puzzled by a piece of reasoning.

'a delightful and engaging introduction to the basic concepts of logic. Whilst not shirking the problems, Priest always manages to keep his discussion accessible and instructive.'

Adrian Moore, St Hugh's College, Oxford

'an excellent way to whet the appetite for logic. . . . Even if you read no other book on modern logic but this one, you will come away with a deeper and broader grasp of the *raison d'être* for logic.'

Chris Mortensen, University of Adelaide

www.oup.co.uk/isbn/0-19-289320-3

PHILOSOPHY
A Very Short Introduction
Edward Craig

This lively and engaging book is the ideal introduction for anyone who has ever been puzzled by what philosophy is or what it is for.

Edward Craig argues that philosophy is not an activity from another planet: learning about it is just a matter of broadening and deepening what most of us do already. He shows that philosophy is no mere intellectual pastime: thinkers such as Plato, Buddhist writers, Descartes, Hobbes, Hume, Hegel, Darwin, Mill and de Beauvoir were responding to real needs and events – much of their work shapes our lives today, and many of their concerns are still ours.

'A vigorous and engaging introduction that speaks to the philosopher in everyone.'

John Cottingham, University of Reading

'addresses many of the central philosophical questions in an engaging and thought-provoking style ... Edward Craig is already famous as the editor of the best long work on philosophy (the Routledge Encyclopedia); now he deserves to become even better known as the author of one of the best short ones.'

Nigel Warburton, The Open University

www.oup.co.uk/isbn/0-19-285421-6